Advanced Sciences and Technologies for Security Applications

Indexed by SCOPUS

The series Advanced Sciences and Technologies for Security Applications comprises interdisciplinary research covering the theory, foundations and domain-specific topics pertaining to security. Publications within the series are peer-reviewed monographs and edited works in the areas of:

- biological and chemical threat recognition and detection (e.g., biosensors, aerosols, forensics)
- crisis and disaster management
- terrorism
- cyber security and secure information systems (e.g., encryption, optical and photonic systems)
- traditional and non-traditional security
- energy, food and resource security
- economic security and securitization (including associated infrastructures)
- transnational crime
- human security and health security
- social, political and psychological aspects of security
- recognition and identification (e.g., optical imaging, biometrics, authentication and verification)
- smart surveillance systems
- applications of theoretical frameworks and methodologies (e.g., grounded theory, complexity, network sciences, modelling and simulation)

Together, the high-quality contributions to this series provide a cross-disciplinary overview of forefront research endeavours aiming to make the world a safer place.

The editors encourage prospective authors to correspond with them in advance of submitting a manuscript. Submission of manuscripts should be made to the Editor-in-Chief or one of the Editors.

More information about this series at http://www.springer.com/series/5540

Zubair Md. Fadlullah • Al-Sakib Khan Pathan
Editors

Combating Security Challenges in the Age of Big Data

Powered by State-of-the-Art Artificial Intelligence Techniques

Springer

Editors
Zubair Md. Fadlullah
ATAC
Lakehead University
Thunder Bay, ON, Canada

Al-Sakib Khan Pathan
Department of Computer Science
and Engineering
Independent University, Bangladesh (IUB)
Dhaka, Bangladesh

ISSN 1613-5113 ISSN 2363-9466 (electronic)
Advanced Sciences and Technologies for Security Applications
ISBN 978-3-030-35641-5 ISBN 978-3-030-35642-2 (eBook)
https://doi.org/10.1007/978-3-030-35642-2

This Springer imprint is published by the registered company Springer Nature Switzerland AG.
The registered company address is: Gewerbestrasse 11, 6330 Cham, Switzerland

"Family and friends" –
Zubair Md. Fadlullah
"My family" – Al-Sakib Khan Pathan

Preface

This book presents a unique opportunity to discuss the development of the Internet of Things (IoT) in the age of big data and its impact on security and privacy of the users. The incentive for gathering, exchanging, and analyzing big data is no longer in doubt for businesses and ordinary users alike. Technology giants like Google, Microsoft, Amazon, Facebook, and Apple and companies like Uber, Airbnb, Nvidia, Expedia, and so forth are continuing to explore new ways to collect and analyze big data to provide their customers with interactive services and new experiences. Cyber-physical systems and IoT are already connecting millions of machines and things, ranging from household devices and industrial appliances to personal wearables. Soon, this number is expected to reach billions as IoT applications like smart home, smart industry, smart city, smart grid, autonomous driving, smart farming, and smart health continue to flourish in an unprecedented way. These applications are generating a massive amount of data which need to be collected and carried by communication networks from the source of generation to analytic sites like cloud servers. The data surge phenomenon has been defined as the big data, which is already having its toll on the existing network systems. The cellular and mobile broadband networks carry the bulk of the user-generated traffic now. Additional pressure from the big data from the aforementioned IoT systems will expose their vulnerability in terms of network congestion and security.

Indeed, with any discussion of big data, security is not far behind. Large-scale data breaches and privacy leaks at governmental and financial institutions, social platforms, power grids, and so forth are on the rise that cost billions of dollars. This book addresses the key security challenges in the big data-centric network and computing systems and discusses how to tackle them using a mix of conventional and state-of-the-art techniques.

The first chapter of the book discusses how big data, despite its ability to change our life for the better, has a detrimental effect on the computing environment. The chapter identifies efficient collection, processing, analysis, and secure storage of the big data. It discusses the security issues concerning big data containing highly sensitive and confidential information. Then, the chapter discloses the trust management techniques prevailing in the IoT with a specific focus on the big data technologies. It sheds light on the trust evaluation foundations, metrics, and evaluation techniques including machine learning, graph theory, game theory, bio-inspired techniques, and probabilistic theories.

The second chapter discusses the concept drift for big data which refers to a change in the statistical distribution of the data. Existing learned models suffer from predictive accuracy loss due to the concept drift. This chapter will serve as a reference to academicians and industry practitioners who are interested in the niche area of handling concept drift for big data applications.

The third chapter proposes a novel taxonomy which groups the classification of outlier/anomaly detection methods based on either quantitative or semantic learning. The viability of the proposed definition of semantic learning is demonstrated in the detection of credit card frauds.

The fourth chapter explains how the increase of digital communication systems increases the potential of side-channel attacks like differential power analysis and correlation power analysis. To prevent these attacks, a cognitive countermeasure is proposed in the chapter by altering the measured power consumption in affecting the secret key value of the power analysis. Hardware prototype-based results show how this proposal can thwart the side-channel attacks.

In the fifth chapter, the mobile ad hoc network (MANET) is revisited to show how it is relevant to the big data scenario. MANETs can be the delivery network for collecting and delivering the big data collected from IoT systems. Collecting reliable data securely in the MANET environment is a challenge due to selfish behavior and trust issues in such distributed network topologies. The chapter addresses these issues in a comprehensive manner by presenting a survey on the secure routing protocols in MANET.

The following chapter surveys the bio-inspired deep learning methodologies and mainly concentrates on the deep learning techniques aiding in the authentication feature extraction process and detection of threatening invasions and malware. It neatly summarizes the challenges for developing the algorithms for IoT networks suitable for diverse application scenarios.

The seventh chapter overviews existing malware detection methods such as behavior-based and specification-based techniques. It also revisits the machine learning-based malware detection methods. Then, a new method leveraging deep learning-based malware detection is presented for IoT-based biomedical systems as well as cloud platforms.

The eighth chapter brings an interesting twist to the book by discussing how blockchain can improve the security of big data by strengthening the security of the data storage. It also explains how this technology can enhance the performance of big data analytics by providing a better data veracity and ID decentralization.

The last three chapters are dedicated to securing the smart grid, which is often regarded as a practical IoT use case. In the ninth chapter, a lightweight authentication method to secure machine-to-machine communication in the smart grid is adopted. A special scenario involving secure targeted broadcast in the smart grid is also presented. In the penultimate chapter, practical defense and forecasting approaches for combating malicious intrusions in the smart grid are discussed.

In the final chapter, an efficient distributed multicast key management scheme, inspired by the blockchain technology, is proposed to provide secure interaction among the smart grid users without the need of a trusted intermediary.

In summary, on the one hand, the book sheds light on how conventional security provisioning techniques like authentication and encryption need to scale well with all the stages of the big data-centric system to effectively combat security threats and vulnerabilities. On the other hand, the book uncovers the state-of-the-art technologies like deep learning and blockchain, which can dramatically change the security landscape in the big data era.

Thunder Bay, ON, Canada — Zubair Md. Fadlullah
Dhaka, Bangladesh — Al-Sakib Khan Pathan

Acknowledgment

We sincerely thank all the authors who contributed their valuable chapters to this book. Thanks to the editorial staffs who helped us in every step of completing this project. Special thanks to our own family members who constantly support our professional work that may often hamper personal life issues. And, finally, we express our sincere gratitude to the Almighty who has allowed us the time and capability to complete this project.

Contents

Contributors

Adang Suwandi Ahmad Cognitive Artificial Intelligence Research Group (CAIRG), School of Electrical Engineering and Informatics, Institut Teknologi Bandung, Bandung, West Java, Indonesia

Mohiuddin Ahmed Lecturer of Computing and Security, School of Science, Academic Centre of Cyber Security Excellence (ACCSE), Edith Cowan University, Joondalup, WA, Australia

Mohamed Baza Department of Electrical and Computer Engineering, Tennessee Tech University, Cookeville, TN, USA

Biozid Bostami Islamic University of Technology, Dhaka, Bangladesh

Pushpita Chatterjee Old Dominion University, Norfolk, VA, USA

Adly Tag Eldien Department of Electrical Engineering, Faculty of Engineering at Shoubra, Benha University, Benha, Egypt

A. K. Fabi Indian Institute of Information Technology and Management – Kerala (IIITM-K), Trivandrum, Kerala, India

Zubair Md. Fadlullah Computer Science Department, Lakehead University, Thunder Bay, ON, Canada
Thunder Bay Regional Health Research Institute, Thunder Bay, ON, Canada

Mostafa M. Fouda College of Engineering, Tennessee Tech University, Cookeville, TN, USA
Faculty of Engineering at Shoubra, Benha University, Benha, Egypt

Michael Gencarelli IVEY Business School, London, ON, Canada

Uttam Ghosh Vanderbilt University, Nashville, TN, USA

Ahmed Ibrahim Computer Science Department, Western University, London, ON, Canada

Rasha Kashef Department of Electrical, Computer, and Biomedical Engineering, Ryerson University, Toronto, ON, Canada

Yusuf Kurniawan Cognitive Artificial Intelligence Research Group (CAIRG), School of Electrical Engineering and Informatics, Institut Teknologi Bandung, Bandung, West Java, Indonesia

Mohamed Mahmoud Department of Electrical and Computer Engineering, Tennessee Tech University, Cookeville, TN, USA

Hala Mansour Department of Electrical Engineering, Faculty of Engineering at Shoubra, Benha University, Benha, Egypt

Mahmoud Nabil Department of Electrical and Computer Engineering, Tennessee Tech University, Cookeville, TN, USA

Al-Sakib Khan Pathan Department of Computer Science and Engineering, Independent University, Bangladesh (IUB), Dhaka, Bangladesh

Septafiansyah Dwi Putra Informatics Management, Politeknik Negeri Lampung, Kota Bandar Lampung, Lampung, Indonesia

Raihan Seraj Department of Electrical and Computer Engineering, McGill University, Montreal, Canada

Arwin Datumaya Wahyudi Sumari Department of Electrical Engineering, State Polytechnic of Malang, Malang, East Java, Indonesia
Faculty of Defense Technology, Indonesia Peace and Security Center (IPSC), Indonesia Defense University, Sentul, West Java, Indonesia

K. S. Sunitha Krishnan Indian Institute of Information Technology and Management – Kerala (IIITM-K), Kazhakkoottam, Kerala, India
Cochin University of Science and Technology, Kochi, Kerala, India

Sarwono Sutikno Cognitive Artificial Intelligence Research Group (CAIRG), School of Electrical Engineering and Informatics, Institut Teknologi Bandung, Bandung, West Java, Indonesia

Sabu M. Thampi Indian Institute of Information Technology and Management – Kerala (IIITM-K), Trivandrum, Kerala, India

Satriyo Wibowo Indonesia Cyber Security Forum (ICSF), South Jakarta, Indonesia

Secure Big Data Transmission with Trust Management for the Internet of Things (IoT)

A. K. Fabi and Sabu M. Thampi

Abstract Big data and Internet of Things (IoT) are the highly sought-after frameworks these days because IoT connects abundant amount of data which cannot be labeled and stored in the typical database system. Generating intelligent decisions from enormously increasing data in a real-time system is of major concern. Although big data seems to change our lives, it tries to make a burden in the computing environment due to the proliferation of data. In such a context, the efficient collection, processing, analyzing and secure storage are identified to be some of the crucial steps. The continuous flow of incoming data to the big data is the first and major challenge and this concern may play a key role in designing a viable and secure big data. In addition, the security issues can be even worse when the stored data include highly sensitive and confidential information. Therefore, if strong security measures are not applied in big data storage, it will cause some vital consequences. Trust management can be considered as a critical factor which operates seamlessly behind the scenes in IoT big data era to provide a reliable communication between devices. This chapter aims to disclose the trust management techniques prevailing in IoT with a special focus on big data technologies; and will outline the new developments and approaches that are applicable in these areas.

1 Introduction

Living organisms use various methods of communication. As the means of communication went on advancement with the gifted brain of primates, human beings developed verbal means of communication which gained command over gestures and other methods. Recently, the digital revolution has caused fast and easy ways of communication. With the advent of the internet, communication and dissemination

A. K. Fabi · S. M. Thampi (✉)
Indian Institute of Information Technology and Management – Kerala (IIITM-K), Trivandrum, Kerala, India
e-mail: fabi.res17@iiitmk.ac.in; sabu.thampi@iiitmk.ac.in; smthampi@ieee.org; smthampi@acm.org

Z. Md. Fadlullah, A.-S. Khan Pathan (eds.), *Combating Security Challenges in the Age of Big Data*, Advanced Sciences and Technologies for Security Applications, https://doi.org/10.1007/978-3-030-35642-2_1

of data became much faster and easier among people everywhere in the world. Various devices analyze and understand their needs, and communicate with each other using the internet without the real - time intervention of human beings. This is referred to as 'Internet of Things' (IoT). The advancement of IoT is inevitable and highly promising for many activities like healthcare, transportation, rescue services, construction, defense, industry etc. [1]. IoT encompasses a large number of devices, and this rapid growth in the number of devices is directly reflected in the data consumption of the network. The outcome of all the data produced by IoT networks can be considered as the source of Big data [2]. Big data contains large collection of data with different processing speed and variety which is not possible to process in the traditional database management systems. Big data and IoT are two rapidly growing emerging technologies. Big data and IoT can work together to provide a highly efficient environment suitable for several applications.

Although there is much advancement, IoT is prone to several security issues due to the presence of fraudulent objects. In addition, many research challenges such as interoperability, data management, energy management, security, privacy and trust also exist in IoT. Among them, security, privacy and trust are the most sensitive requirements for the success of IoT. The IoT paradigm will not mark a prominent shift without a secure technology that ensures user privacy, safe communication and trustful interaction. However, current literature still lacks a comprehensive study on trust management in IoT. Unquestionably, many security challenges must be addressed in big data in accordance with IoT.

What happens in Big data is the exploitation of the data itself since IoT connects all devices across the globe. In this context, if a security measure is applied in IoT network, the data flow into the Big data can be reduced by increasing the confidentiality of the data in the same. Living beings interact and work together to achieve either one's own goals or a group's goal. But still, individuals are sceptic about the trust put by another one. No one shares anything crucial until complete trust is established. The same is the case with IoT devices too. Existing trust management techniques focus on fuzzy logic, edge computing, deep learning, blockchain and so on to achieve the trust. To date, there is a lack of trust computing mechanisms that ensure all the possible functionalities of IoT environments [3].

This chapter aims to disclose the trust management techniques prevailing in IoT with a special focus on Big data technologies, and will outline the new developments and approaches that are applicable in these areas. Primarily, we provide a classification of various common IoT trust computing techniques. We also aim to uncover the scope of big data processing and analytics platforms in the IoT environment. Finally, we conclude the chapter with the current security challenges and the future research directions of big data analytics in IoT platform.

2 How Will IoT Impact Big Data?

IoT and big data are becoming the most relevant technologies nowadays and these two have the potential to amend many aspects that exist in the network. IoT is a promising technology which will take big data to a different level. When a large number of devices are collaborated to work together in IoT, they will produce a continuous stream of data. Once the data have been collected, the next target is to find the best solution to store these data in a proper manner. Traditional data warehouse systems are very expensive and time-consuming. The cost of processing and storing the large collection of data can be considerably reduced if we use an efficient technology in real-time data analysis. Here comes the need of big storage space to accommodate the data and this can be referred to as big data.

The huge data generated by the IoT will affect the entire big data universe and force it to upgrade current tools, processes and technologies in the same so as to accommodate the additional volume of data. The devices present in the IoT may vary according to the kinds of data they produce and the variety of data produced may demand different kinds of security solutions to protect them from risks. In IoT environment, Big data is used for collecting, processing and storing the data generated by the sensors and other nodes and for making decisions according to the analysis made from the data.

Figure 1 shows the relationship between IoT and big data. A large number of heterogeneous devices connected in the IoT network generate a massive amount of data when they are communicating with each other and would directly impact the big data. Moreover, there exists a real time sharing, analysis and transmission of data.

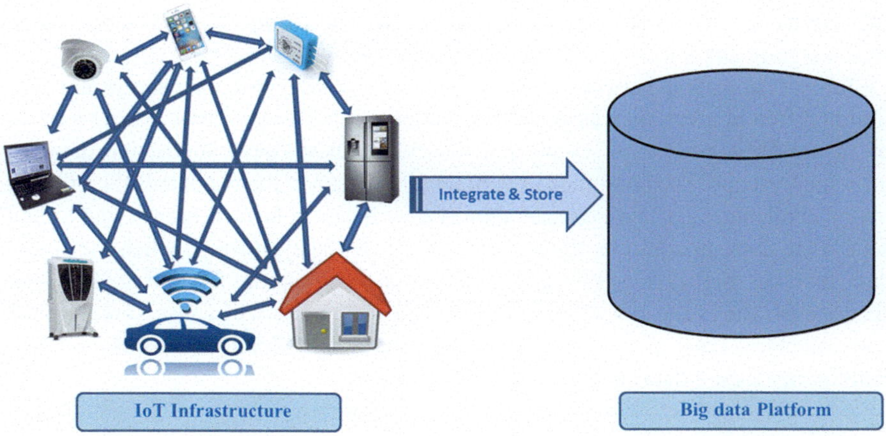

Fig. 1 IoT and big data processing

2.1 IoT Based Big Data Solutions

IoT generates data at a huge and faster rate resulting in an increased demand for the big data storage technology. The combined application of IoT and big data enhances the opportunity of researches in both fields. Nowadays several papers focus on the relationship between IoT and Big data. The key requirement for the management of data and the recent advances in IoT and Big data environment were discussed by Ahmed et al. [2]. Bashir and Gill [4] defined the need of integrated IoT Big data analytics in the Big data environment specifically designed for IoT applications to fulfil the gaps identified in this area. The data generated from 15 virtual IoT sensor nodes were extracted and analyzed in real time environment and some actions like alarming with sound or light were proposed.

Ahmad et al. [5] proposed a big data analytics to predict human behaviour from the real-time data driven by the social internet of things. It facilitates understanding the human behaviour by correlating SIoT and Big data and its applicability was demonstrated using Hadoop ecosystem. This system utilized a feedback system to offer a chance to improve human behaviour. Ilapakurti and Kedari [6] examined the role of big data in electronic health record (EHR) monitoring patient's health from sensors based on IoT framework. They proposed a framework that comprised low powered Bluetooth, embedded sensors, software edge analytics, and multidimensional Big data analytics. Low energy Bluetooth was used to connect sensors and then edge processing was applied to the data captured. It is also capable of visualizing individual patient's health in real time and detecting urgent situations.

Arora et al. [7] discussed the significance of using machine learning based classifier, which can beat the performance of other classifiers when there is a large dataset. Berlian et al. [8] proposed a real - time system framework to monitor the internet of underwater things. Instead of SQL, they used Hadoop MapReduce to process the query which reduced the query execution time. Table 1 shows some recent significant works on IoT based big data solution.

Table 1 shows IoT network security measures based on big data and its application in different areas such as healthcare, industrial IoT, smart cities etc. As a result of fusing large number of connected devices in these applications, the data are growing by leaps. The continued flow of incoming data to the big data is the first and major challenge and this concern should play a key role to design a viable and secure big data. In addition, the security issues can be even worse when the stored data includes highly sensitive and confidential information. Therefore, if strong security measures are not applied in big data storage, it will cause some crucial consequences. Trust management operates seamlessly behind the scenes to endow with highly secure data storage in IoT big data era.

Table 1 Significant works on IoT based Big data solution

No.	Approach	Finding and descriptions
1	IOT-Statistic [9]	Defined a general statistical database cluster mechanism for big data analysis.
		The statistical analysis is carried out in a distributed and parallel manner using Euclidean-based spatial aggregation.
2	Human Behaviours analysed using Big data Analytics [5]	Human behaviour was analysed using big data analytics and SIoT.
		Data were captured from smart cities and wearable devices to define the human behaviour using Big data.
3	The Role of Big data in creating Sense EHR [6]	Created sense EHR based on IoT framework.
		Used low energy Bluetooth.
		Edge processing was applied to filter, aggregate, enrich, and analyze a high throughput of data from the sensors.
4	Big data Analytics for classifying Network Enabled Devices [7]	Classified network enabled devices using Big data analytics.
		Four machine learning algorithms were used and their F-scores were calculated.
		Results showed that machine learning based classifier models provide feasible solution to large dataset.
5	IoT big data analytics in smart building system [4]	Proposed an integrated IoT Big data Analytics framework.
		The components of IBDA were IoT sensors, big data management and analytics.
		PySpark was used for analytics.
6	IoT Based Cyber Physical System for Industrial Big data Analytics [10]	An interchangeable cyber physical system for big data analytics in industrial IoT was proposed.
		The system contained the inclusion of RFID.
7	IoT Framework with Semantics, Big Data, and Analytics [11]	Big data, IoT and semantic web are integrated in the framework.
		Described the necessity of semantics in IoT and big data.
		It will provide an effective way to combine all the sensors to store the data using some semantic rules.
8	Big data and Industrial Internet of Things [12]	Described opportunity and challenges in big data and IoT.
		Proposed a framework that integrates both big data and IIoT.

3 Trustworthiness Measurement and Evaluation for IoT Big Data

A device may be considered a fraudulent or malicious node in the network either because of its intentional fraudulent action or because of the wrong service due to its probable hardware failure. Nevertheless, both of these issues may lead to the disorder in the services of other objects and raise a challenge to the security of private data shared by them. The combination of authentication and access control was considered as a security approach to the distributed network in the early stage of the internet era. But as the number of heterogeneous devices in the world increased within no time, researchers realized that what they followed was not powerful enough to handle the security problems because of the dynamic nature of the devices. This section explains the importance of trust management in IoT big data era. Trust is a condition in which the beneficiary of a service can rely on the service of a provider without being skeptic about any fraudulent activity. For instance, Trustor will make a measurable belief on the trustee for a given task in a specific time period. Here, Trust management is the combination of gathering all the information regarding trust relationship and securing and monitoring all the activities in communication. Figure 2 shows the environment in which trust can be applied which includes formation, distribution, and updating of trust values in each node. Simply, it is a relationship between trustor and trustee and they will yield a trust between them. In this study, we are focusing on trust management systems that are applicable to the IoT big data era. Even within the same network domain, trustworthiness of a node may change hence there should be a consistent trust management system in the network.

Initially, Blaze et al. [13] introduced the term trust management in their work in which they defined trust management as the unified approach which contains security policies, credential and relationship that allow direct authorization of security-critical actions. Over the past few years, many works have focused on Trust management in several network environments, such as social networks, peer to peers (P2P), grids, adhoc networks, clouds, edges and IoT.

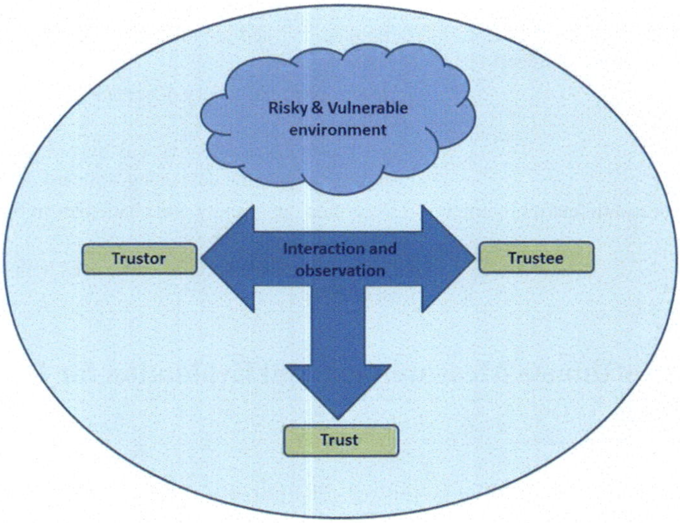

Fig. 2 Trust management environment

Trust management becomes more challenging as the network infrastructure changes from a fixed network topology in to mobile network. The mobility of nodes and the lack of permanent infrastructure make Mobile Adhoc Network (MANET) and Wireless Sensor Network (WSN) more vulnerable to attacks. So it is very mandatory to provide a trust management to these networks. WSN is the special type of mobile ad-hoc network which contains independent sensor nodes without any central server system. WSN is the collection of sensor nodes which monitor and collect the physical characteristics of environment like temperature, humidity, weather etc. So it can be used in different areas concerned with military, agriculture, industry and civilian. WSN can be used in short distance applications, for instance, it can be used as a combination of other networks like IoT. Another emerging paradigm of IoT is the social internet of things (SIoT) where the things owned by person can also be the users of social networks. Social network users share their feelings, photos and other sensitive data indiscriminately in social networking sites. Due to the transparent nature of social networks, most of the existing trust management systems propose a framework which separates the two layers that consists of people and things [14–16].

3.1 Motivation to Trust Management

The tremendous volume of data generated incrementally leads to the creation of so many security and privacy issues in big data. Researchers have already addressed several techniques to overcome such challenges over the past decades. For this scenario, cryptographic techniques, trust management, access control, firewall, data guard and data provenance are some of the important techniques that are used to secure big data. In order to protect data, all the traditional methods such as firewall and encryption use two factor authentications. Cryptographic techniques won't fit well with the big data IoT ecosystem which is composed of billions of extremely heterogeneous devices with different capabilities.

However there are some points that should be kept in mind while constructing security measures in an IoT network,

(a) No secured items are more secure than the threat modelling done on it.
(b) Every secured item is sensitive to something.
(c) The perfection of any security has got a limit as no one knows all probable risks or attack vectors.

So, it is essential to understand and quantify the level of threat and vulnerabilities existing in the IoT network before implementing any security solution. Otherwise, there may be a mismatch between threat and chosen security measures leading to performance degradation, service denial and data compromise [17]. This mismatch is often found in security systems where cryptography is deployed. Also, if an intruder entered into a network that uses encryption techniques, the intruder can directly access plaintext or it can modify the content or protocol in order to get access. Hence, any security solution that we select should satisfy all the current and future perspectives of the network such as storage and transport characteristics of data. Another fundamental aspect of data security is to provide selective access control and defend the threats like denial of service which cannot be satisfied using cryptography.

With the growth of the IoT, trust management is being more and more prevalent. The current research in trust management in IOT is highly motivated by some of its major advantages, which are summarized as follows [18].

(a) Trust management allows access control to the services and resources using some credentials and make a trust chain for the propagation of access right.
(b) It is more expressive and scalable than any other classical security system such as encryption and access control mechanism.
(c) It is not a global concept: trust is established in one particular device and just two parties are involved in the trust establishment.

3.2 Types of Trust Management Applied in Different IoT Networks (Fig. 3)

3.2.1 Centralized

Flexibility and heterogeneity are the major challenges encountered when IoT deals with large number of devices in the network. In this situation, IoT security challenges can be overcome using centralized trust management mechanism where centralized entity is either a participating IoT device or a physical cloud. The centralized device keeps the table that contains the trust information of all devices participating in IoT network. A novel centralized trust management model was proposed in [19] to provide a trustworthy interaction between IoT devices where a super node serves as a central trust manager all over the network. Whereas, a centralized trust management in social IoT was proposed in [20] where the trustworthiness of each device extracted on the basis of its own experience and the opinion from common friends was discussed. On the basis of these observations, a distributed hash table was constructed to store the trust values of all devices. A generic trust management system has to be made in the IoT to ensure the reliable communication among devices

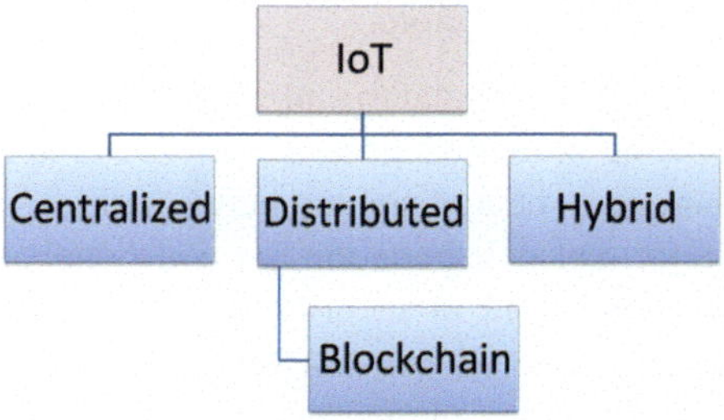

Fig. 3 Trust Management applied in different IoT networks

according to the different context (status of the nodes) and different functions (for which service) in the network. A context aware approach was proposed in [21] using a central trust information of IoT devices.

3.2.2 Distributed

In centralized trust management system, the central authority may be attacked by hackers. Once the intruders access the central system, massive amount of data would be stolen and they gain control over other activities on sensitive data. Moreover, centralized trust management systems cannot give enough trust due to the presence of intermediates in between data exchange. As a remedy, distributed trust management mechanisms can be deployed without the real time intervention of a third party during transactions. Similarly, distributed systems can reduce the high cost and difficulty incurred in data storage more than the centralized trust management systems. Literature [22] includes a distributed computing in trust management detecting trustless devices in the system employing fuzzy sets. Similarly, [23] discusses a distributed collaborative filtering for the selection of trustworthy devices using friendship rating and social contact.

3.2.3 Blockchain

Blockchain systems have been recommended as an efficient trust management mechanism for distributed environments. Further, distributed trust mechanism is leveraged by using blockchain within it as it brings new potential to the IoT-Big data era. The blockchain has the potential to secure the data and provide the access to only the trusted parties along with a credit value. Once entered, data can never be removed since the transactions that are present in blockchain ledger were accepted and verified by majority of the devices. If a device misbehaves in a particular context, its further operations can be prevented since the read only nature of the ledger makes it impossible to carry out any fraudulent actions.

3.2.4 Hybrid

In fact, blockchain model in IoT can overthrow different attacks in networks. Although blockchain technology in Trust management of IoT brings high level expressions of security, privacy, storage and computing, cost is one of the major concerns. To solve these problems, some IoT networks use a hybrid network instead of single centralized or distributed network. A trust management protocol developed for hybrid IoT environment was discussed in [24] using hierarchical cloud architecture. This model was based on social similarity and collaborative filtering. Similarly a hybrid trust management to filter out malicious data and data sources is implemented in [25] using learning algorithms.

4 Trust Evaluation Foundations

4.1 Trust Concepts

Many researchers talk clearly and simply about the term Internet of things (IoT) though it is an understandable one with a single definition. However, specifying the IoT is not as easy as it encompasses wide range of technologies, processes and applications. According to ITU [26], IoT is defined as a global infrastructure used for the information society, enabling advanced services of interconnecting things on the basis of existing and evolving interoperable information and technologies.

4.2 Trust Metrics

There is no a ratified way to distinguish the level of trust. With each depiction of a trust value, researchers considered multiple aspects of device's behavior and this can be referred to as trust metrics. Different methods are used to represent the trust metric. Certain networks assume only binary values such as 0 and 1 where 0 is used to represent distrusted and 1 for trusted devices, or even comprise some intermediate states using the values in between 0 and 1. Whereas in some networks they use some fixed scale, or some range of values. Consequently explicit values can be labeled to the trust that is evaluated to each node in the network. In some trust management protocol, a node manages a trust metric which takes three parameters for the evaluation of trust. In dynamic trust management protocol discussed in [27], a node maintains a trust metric with multiple properties like cooperativeness, community-interest and honesty. For instance, in [28] experience, knowledge and recommendation are maintained as trust metrics. Apart from these, most of the researchers use reputation based metrics for the evaluation of trust [29, 30]. Reputation expresses the estimation of reliability through the social control. Such metrics of network is shown in the Fig. 4.

Evaluated trust value was compared with a predefined threshold level. According to the obtained trust level, additional rewards were given to each node [31]. Similarly, some of the major trust management methods use a punishment or penalty value for the malfunctioning and untrustworthy nodes [31]. Table 2 summarizes the different trust metrics used in IoT network.

5 Trust Evaluation Techniques

Direct trust and indirect trust are mainly used for the trust calculation. Direct trust is formed between the connected nodes and indirect trust is formed between the longest connections of nodes that are having intermediate nodes. If we are unaware

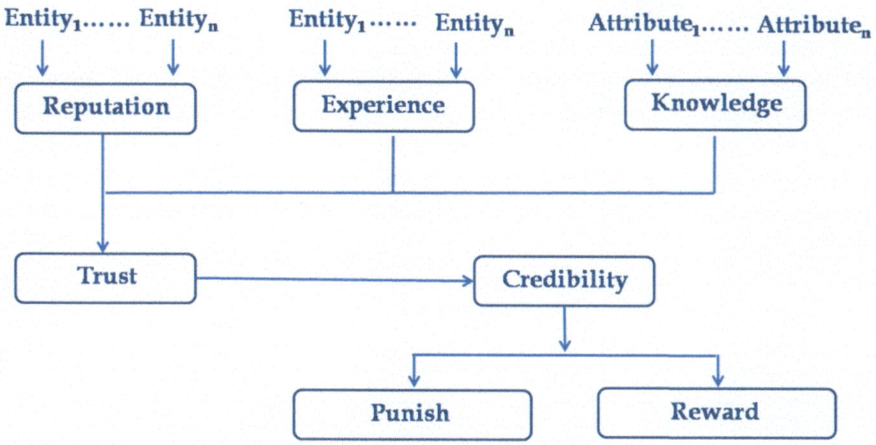

Fig. 4 Trust metrics with attributes reputation, experience, and knowledge

Table 2 Summary of different trust model with the metric used in IoT network

Approaches	Metric employed
A context-aware trust management [21]	Capability
	Service
Dynamic Trust Management [32]	Honesty
	Community of interest
	Recommendations
	Cooperativeness
A fuzzy approach to trust management [33]	Recommendations
	Knowledge
	Experiences
Hierarchical Trust Management [34]	Intimacy
	Honesty

of a particular node from which we need some service, we seek recommendation from those which are familiar with them. The recommendation system provides an improvement over the traditional direct evaluation methods [35].

5.1 Machine Learning

In machine learning systems are made able to learn automatically and improve from experience, in order to solve a specific problem and to predict factual results. Different trust features can be effectively processed to evaluate trust management framework using machine learning. Numerically measurable values are generated in [36] using the combination of machine learning and mathematical methods which produce a final trust value from raw data. Here, extracted trust features are labelled using clustering algorithm and a multiclass classification algorithm (see Fig. 5) like support vector machine is used to combine all measured trust values. There are some other works [37] which also represent a support vector machine for multiclass classification problems. Instead of using direct observation values, it uses a feedback value from the neighbouring nodes. It also uses an n-fold cross validation set to predict the node's trust level. Table 3 describes the major researches on machine learning algorithms.

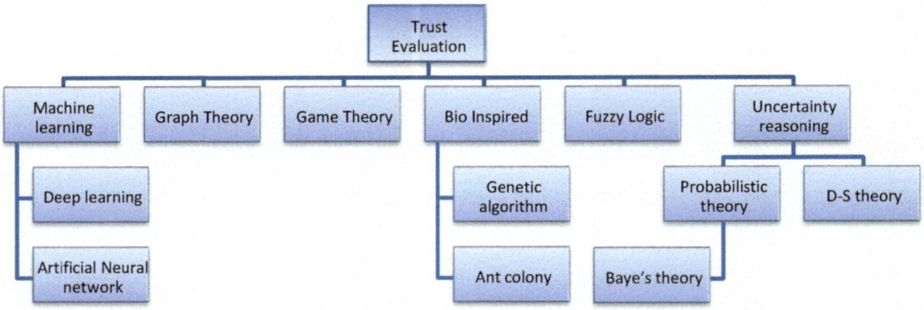

Fig. 5 Classification of trust evaluation techniques

Table 3 Significant IoT trust management works based on machine learning

Approaches	Method and inferences
Machine Learning based Trust Computational Model [36]	Combining machine learning and mathematical methods.
	Unsupervised learning is used to label trustworthy and untrustworthy nodes.
	Extracted trust features are labelled using clustering algorithm.
	Support vector machine is used to combine the evaluated trust values.
Trust Management based on (SVM) support vector machine [37]	SVM based trust model using Gaussian kernel.
	Trust evaluation done by RESTful web services.
(ANFIS) Artificial neural fuzzy inference system [38]	Brain inspired trust management model.
	Utilized weighted-additive method to combine both behavioural and data trust.
	Three linguistic terms were given as input to the system and then nineteen possible 'If then else rules' were formulated.
Trust Management Method Detecting On-Off Attacks [39]	Using limited amount of data to identify on- off trust attack.
	Attacker nodes and broken nodes are separated.
	Combined both machine learning and elastic slide window.
	Executed in both simulated and real world data.

5.1.1 Artificial Neural Network

The Brain is the most important organ in the human body and it helps to think to understand and take the decision, and the secret behind all the power of these is neuron., Scientists have been trying to mimic the basic functions of brain neurons to build robots since 1950. As the name suggests, Artificial Neural Networks (ANN) are inspired by the human brain. In the artificial neural network, a large number of hidden layers are present, which is formed of a set of neurons. A brain inspired trust management framework is introduced by Mahmud et al. [38] in which a fuzzy inference system is used to evaluate the trust in IoT. They mentioned 19 possible inference rules to predict the node's behaviour to evaluate the trustworthiness of each node and utilized weighted-additive method to combine both behavioral and data trust. Whereas, Caminha et al. [39] proposed an elastic slide window technique which is used to detect broken or malfunctioning nodes. Hence, this paper is mainly concerned for detecting on-off attacks.

5.1.2 Deep Learning

Deep learning is a machine learning technique that attempt to learn multiple levels of representation by using a hierarchy of multiple layers. If we are giving huge amount of data it begins to understand it and responds in useful ways. Deep learning provides a very flexible, universal, learnable framework for representing visual and linguistic information. It can learn both supervised and unsupervised data. Combination of deep learning with matrix factorization was implemented in [40] using a recommendation problem in social networks.

5.2 Graph Theory

The graphical approach is mainly used for network flow analysis and routing purposes and it uses geometric structures (graphs) to model relationship between devices. In general, an IoT network can be represented as G = (V,E) where V represents the individual devices and E represents the trust relationship between devices. Numerous researches have proposed graph theory-based approaches for improving trust among nodes. A very common method in graph-based trust management is the random walk method, in which transition probability is assigned by walking through each edge [41]. Jiang and Barashave discussed [42] the trust graph management through a local voting system and they establish a topology using algebraic graph theory to spread the trust value to all over the network. More significantly, this approach can be used in real time communication to abstract the trust data. Many of the researchers concentrated on graph similarities problem [24, 25]. Mainly, two types of similarities were defined, viz., Trustee based similarity and Trustor based similarity [43]. k-most similar trustees were determined in the first method and

Table 4 Significant IoT trust management works based on graph theory

Approaches	Method and inferences
Trust using subjective logic [35]	Weighted directed acyclic graph is used.
	Uncertainty is also considered.
	Opinion triangle is used where vertices represents belief, disbelief and uncertainty.
	Opinion is taken from user's binary observation towards each node.
Graph Algebraic Interpretation [42]	Established a voting system in the network for trust evaluation rule using markove chain.
Eigen trust algorithm [45]	Global trust value was calculated for each peer.
	Neighbours opinions are weighted and normalized to evaluate the trust value.
	Minimize the impacts of malicious peers.
Resilient routing mechanism [46]	Proposed a new light weight RPL protocol.
	Node's reputation values were calculated.
A Graph-based Trust-enhanced Recommender System [47]	Without any central system each node can calculate its own prediction.
	Formulate an algorithm to measure the impact of one node to another.

k-most similar trustors were identified in the second method for computing similarities. Apart from this, Melnik et al. [24] worked on the similarity problem called similarity flooding, where the final steps need to be adjusted by humans. Another approach for addressing trust using graphic theory was proposed by Khan et al. in [26] using trust-based routing mechanism where the reputation value of each node in the IoT networks is mainly focused. Regarding the graph theory based trust, some of the trust management systems either do not take trust relationship factor or context dependency into account [25] or predefine. Gemini et al. [44] investigated a graphical security framework to secure the IIoT system from vulnerability-based attacks by selectively removing the high-risk attack paths. A subjective logic is used to model in [18].

Trust can be visualized to convey the relationship among the nodes [27] hence we can handle with desired outcomes. In addition, graph theory provides a good method to study the insightful behaviors and interactions of devices in IoT networks. Another method [42] based on markove chain establishes a voting system in the network for trust evaluation rule. Table 4 explains some of the major approaches in graph theory.

5.3 *Game Theory*

It is a competition theory that formed between two or more decision makers to make a decision according to their own benefits. Some of the researchers considered Bayesian game [48, 49] where both players might know their own strategies and

Table 5 Significant IoT trust management works based on game theory

Approaches	Method and inferences
Bayesian Game in Trust Management [48]	Proposed a Bayesian game between users and trust devices.
	Two players reach a Nash equilibrium state.
An incentive mechanism based on game theory [49]	Developed a Bayesian game between malicious unknown nodes and normal nodes.
	Trust is ensured by encouraging the nodes to cooperate by giving incentives.
A Game-Theoretic Approach for Data Trustworthiness [50]	Uses a discrete time model.
	Play a Stackelberg game and reaches a Nash equilibrium condition.
Game theory in fuzzy large-scale networks [51]	Initially proposed a fuzzy decision making model.
	Multi-criteria fuzzy decision-making (MFDM) model was proposed for trust prediction in fuzzy large-scale networks.

payoffs. Here user and trust domain are the players of the game. Finally they will reach an equilibrium state which shows the best reaction of one player to the plan of action of another player. A defence mechanism in wireless sensor network based on game theory was depicted in [50] where, Trustor and trustee perform a Stackelberg game and they will reach a Nash equilibrium condition. Whereas Fang et al. proposed [51] a linguistic fuzzy model in game theory. Table 5 explains some of the major approaches based on game theory.

5.4 *Bio Inspired*

Human beings always find a solution to any problem from nature and execute them for solving many computing problems. Algorithms developed out of the inspiration from nature are called bio-inspired or nature-inspired algorithms. In this view, biological activity can also be used to model trust and it has been employed as a reliable technique in trust management. Many researchers developed bio-inspired trust models influenced by ant colony. Some researchers followed the approaches that evaluate trust using particle swarm [52] or a genetic algorithm.

5.4.1 Ant Colony Algorithm

Most commonly used bio-inspired technique in trust management is the Ant Colony Algorithm (ACA). Even though there are many bio-inspired approaches for trust management in distributed networks, ACA proves to be one of the best approaches for adapting the capabilities in highly constrained and autonomous environment. The biological facts behind the algorithm are the following. Ants scout randomly in search of food nearby their nests. Once the food is spotted, the scouting ant takes a sample and return home discharging a trail pheromone on the path. Sensing this smell, the forager ants follow the trail for collecting the food. If there are many paths with trail pheromone, the one with more concentration of pheromone will be chosen. The path selected by more ants will become stronger as each secretes trail pheromone and the unselected ones will gradually disappear. The quality and quantity of the food determine the concentration of pheromones secreted. Similarly, the optimum path can be easily adjusted accurately to the trustworthiness of peers and the availability of network resources.

5.4.2 Particle Swarm Optimization

Particle swarm optimization (PSO) algorithm is inspired from the behavior of bird flocks and schooling behavior of fishes in nature. Flock of animals or birds follow certain patterns through continuous iteration in order to achieve the finest position of each individual and this movement of birds is influenced by their own past locations. In trust management, PSO can be effectively used to select the trusted path to the server or services. A PSO based trust path selection approach is presented in [52] where the equivalent nodes are deleted to scale the complexity of network structure. Mutation is also applied to PSO algorithm [53] to minimize the running time of the same.

5.4.3 Genetic Algorithm

Genetic algorithm mainly concentrates on network quality measures such as energy efficiency, throughput and delay when it deals with trust management. The biological concept of natural selection is the major idea behind the genetic algorithm where fitness function values can be used to get the best population in the next generation.

The first study related to genetic programming on trust named GenTrust was proposed by Tahta et al. [54]. GenTrust constructs GA trees for each individual node using the features acquired from past interactions and recommendations. This study evaluates the trust value of each tree using a mathematical function followed by the exclusion of nodes with lower trust values. Another interesting research on GA was proposed in [55] which detects the most trustful node using traffic analyzer by observing the behavior, timing and optimality of each node. Table 6 explains some of the significant approaches based on nature inspired algorithms.

There are also several other nature inspired approaches which take inspiration from different biological structure found in nature. Even though there are many bio-inspired approaches, ACA, PSO and GA are the most suitable for the trust management in IoT.

Table 6 Significant IoT trust management works based on bio-inspired

Approaches	Method and inferences
Genetic Algorithm Based Trust [55]	A traffic analyzer is used to get the most trusted nodes in the network.
	Observes the behaviour, timing and optimality of each node to ensure the trustworthiness.
Trust calculation using ACO [56]	Calculate trust path and cycle using probabilistic trust rule.
	Using probabilistic trust rule, trust value is calculated from User Activity Matrix.
Heuristic algorithms in IoT [57]	Heuristic methods like genetic algorithm (GA), ant colony optimization (ACO) and particle swarm optimisation (PSO) are opted to get optimal solution.
	Experimental evaluation proves that PSO outperformed all other methods.
Ant Colony System on trust (TACS) [58]	TACS algorithms were executed and optimal path of the most trustworthy server was fetched.
	Pheromone traces are analyzed with the trust value.

5.5 *Fuzzy Logic*

There are some cases where the concept may be of partial truth in the way that the truth ranges between exactly true and exactly false. Moreover, when we are using linguistic variable instead of some binary values, these degrees can be effectively managed. In such cases, fuzzy logic can be effectively useful to handle several types of ambiguity in which some statement would be true or false to some extent. eg: "'a' is a member of 'A'", where the statement maybe either exactly true or exactly false. Hence degrees of trust can be represented as a fuzzy value with membership function such as very low trust, medium trust, and high trust. Complete trust, Ignorance and Distrust are used as the linguistic variable in [59] where 1 implies to the complete trust and 0 to the distrust. Whereas, Mahalle et al. considered good, average and bad as linguistic variable [33] in their work and they assigned membership value to Knowledge, Experience and Recommendation as input and trust is taken as output. Fuzzy logic system uses some scores as the input variable and processes through some fuzzy inference rule generating output variable which can be termed node's trust level. Here, the inference rules are generated from the human experiences and it computes the trust value from the belief sources. Besides, it is abstractly very easy to understand based on subjective logic. Table 7 shows the significant trust management works on fuzzy logic.

Table 7 Significant IoT trust management works based on fuzzy logic

Approaches	Method and inferences
(TRM-IoT) A Trust Management Model in IoT based on Fuzzy Reputation [22]	Package forwarding behaviour of each neighbour was monitored.
	Each node keeps data forwarding transaction table containing behavioural data.
(FTBAC) Trust Based Access Control based on Fuzzy approach [33]	Linguistic information is used to deal with the access control in the IoT.
	Scalable and energy efficient framework is proposed.
A Fuzzy-based Trust Management [60]	Three trust metrics are used including probability of successful interactions, feedback of messages exchanging and battery energy.
	Each node ranks its neighbours to distinguish the selfish node.

5.6 Uncertainty Reasoning

When we are dealing with real time applications, we may have to handle some imperfect information too. In these cases, we can use uncertainty reasoning techniques. In fact, we can measure everything, in our view, even though evaluation metric does not contain complete information due to any one of the following case,

1. Information may be user's assumption.
2. Information may well be inaccessible.
3. Knowledge may be missing.
4. Given data may be unreliable or ambiguous.
5. Data could not be acceptable.

Three approaches for handling uncertainty are:

1. Probabilistic theory
2. Baye's Theory
3. Dampster-Shafer (D-S) Theory

5.6.1 Probabilistic Theory

Regarding the evaluation of trust, some trust management systems takes both probability and uncertainty in to account [61].

Baye's Theory

Bayesian theory is the most suitable theory in the case of uncertainty reasoning especially in updating believes and decision making.

Simply, Baye's formula can be expressed as following.

$$P(A|B) = \frac{P(B|A)P(A)}{P(B|A)P(A) + P(B|A^c)P(A^c)}$$

Here event 'B' has already occurred while event 'A' is upcoming. Trust was considered as a random variable in Baye's theory, so that probability distribution can be simply applied by adding new parameters that were updated during the observation. A Bayesian theory based trust management approach for predicting future trust behavior of a node using a large number of positive and negative experiences was modeled in [62]. Method highlights that the feedback from multiple nodes could be combined to get the average trust value. Another important work focused on the Bayesian inference rule was proposed in [63] where reputation scores were calculated from different positive and negative ratings given by other nodes. Using Baye's theorem following formula was created to get the trust value.

$$P(Belief|Observation) = \frac{P(Observation|Belief) \times P(Belief)}{Normalization}$$

5.6.2 Dampster-Shafer (D-S) Theory

Dampster-Shafer theory is the most commonly used evidence theory in trust management hence called the mathematical theory of evidence. Evidence theory was first proposed by Dempster and then further developed by Shafer. So it was also called Dampster-Shafer theory (DST) or evidence theory [41, 42]. Dampster had proposed a theory to combine both upper and lower probabilities that are generated from independent observations [43]. Moreover, Evidence theory has many advantages over other uncertainty theorems due to the capability of gathering feedback generated from the networks. Neither it will directly express or evaluate the uncertainty, but it shrinks the assumptions that are accumulated by the evidence [64]. In trust management, evidence can be the set of trust information and parameters generated during the interaction of devices. A trust model proposed by [65] implements a bayesian and entropy based trust model, where entropy theory can efficiently distribute weights to various trust values. An algorithm for trusted routing based on the combination of DS evidence and ant colony method was proposed in [64]. Xiang Qiu et al. [66] proposed a D-S trust transitivity model for transforming the triple of evidence theory by using nearness degree. Table 8 explains some of the major approaches based on uncertainty reasoning.

Table 9 presents some of the major approaches in the trust management techniques in the IoT networks and defines basic methodology inferred from it.

Table 8 Significant IoT trust management works based on uncertainty reasoning

Approaches	Method and inferences
A Probabilistic Trust Based on Evidence [61]	Proposed a probabilistic approach for updating the trust values.
	Amount of trust is updated by an agent.
	Trust worthiness provides accurate result where agents change frequently its behaviour.
The Beta Reputation System [62]	Feedback from different users was combined to get the reputation rating.
	Applied beta distribution to the observed reputation values.
Reputation-based Framework [63]	Reputation values are observed through a Watchdog Mechanism.
	After applying beta distribution, Bayesian formulation was used for reputation integration and updates.
Bayesian and entropy based trust management [65]	History records are maintained instead of storing all information in node's memory.
A Trust Transitivity Model Based-on D-S Theory [66]	Two types of trust relationship are identified: identity trust and behaviour trust.
A trust evaluation algorithm [67]	Obtained membership degree using fuzziness theory.

Table 9 Different approaches for trust management in IoT

Approaches	Basic methodology
Ant colony	To obtain the optimal path to the best server or service.
Genetic algorithm	Genetic algorithm finds the optimal solution by maximising or minimising the fitness function.
Bayesian theory	To predict future trust behaviour of a node using a large number of positive and negative experiences.
Game theory	Shows how will the device trust or distrust the other node using different game-plan.
	Derives explicit equilibrium criteria to examine the optimality of the outcomes.
Graph theory	Visualizes information implies level of trust between devices.
Fuzzy logic	Fuzzy logic was used to express the uncertainty in the trust network using the subjective logic of the devices.

6 Block Chain Promising the Trust in the Future IoT Systems

Decentralized trust domain is the major challenge existing in the big data environment. With the powerful background of crypto currencies like bitcoin, a significant solution for the security concerns for IoT based big data can be provided. Blockchain can provide a secure ledger among the untrusted parties that would be expected in the big data environment. Any other trust evaluation technique can be tailored to the blockchain based trust management.

Figure 6 shows the blockchain based transaction where each block represents all the transaction occurred and it is linked to the previous block. So it is a back linked ledger which encodes the transaction occurred in it. Thus, the more the length of the blockchain the more difficult it is to attack the same. Such a trust evaluation platform offers a comprehensive solution for convenient and secure communication among heterogeneous devices.

Figure 7 shows the connection between IoT networks and block chain. All the transactions that happened in IoT network could be stored in each block which is linked to the previous one. Here, any device generating a block needs to provide a solution for a specialized mathematical problem. So it becomes practically impossible for a fraudulent device to add the blocks to the current blockchain. Based on the credit history of a device, another device decides whether a service should be accepted or rejected. Instead of verifying the digital assets kept in each device during run time, obligations are kept in chain which can be locally accepted during transactions.

A blockchain based trust management with a reputation mechanism and access control system is designed in [68] where a credit value was maintained in contrast to bitcoin. Similarly, An obligation chain was used to explain the block chain based

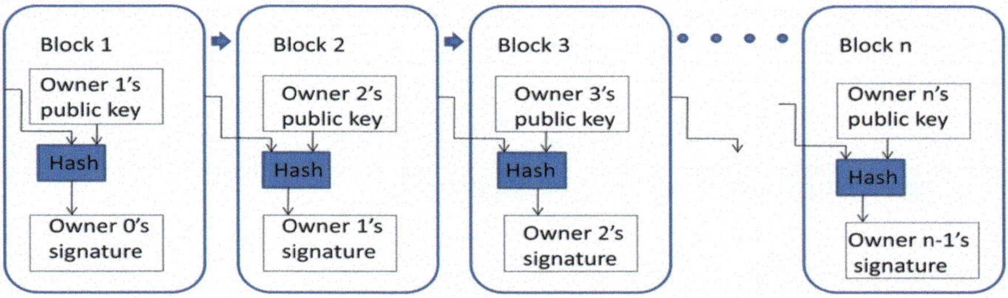

Fig. 6 Block chain based transaction

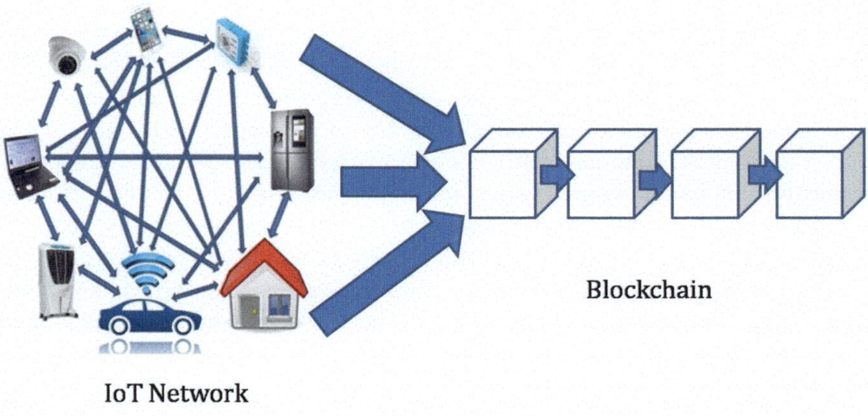

Fig. 7 Trust management done by block chain in IoT network

credit score system in [69] where in trust evaluation is carried out on the basis of a reputation mechanism that also checks the credit history of each peer to provide trustful communication among them. Mohamed et al. introduced a bubble of trust concept [70] where in the bubble was used to represent a virtual zone that ensures the trust of the participants that rely on the blockchain. Even though IoT devices can securely communicate in bubbles of Trust, this mechanism is not suitable for real time applications. An alternative approach was proposed by Zyskind et al. who combined blockchain and off-blockchain storage to implement secured data storage [71]. Juah et al. proposed Hyperledger fabric instead of block chain [72] and implemented the trust management system with the use of Docker Swarm. Apart from this, an Ethereum blockchain based trust mechanism was implemented in [73] where the transaction records are conserved in a transparent and inflexible manner. Similarly, Juan Li et al. [74] proposed a semantics-based mechanism to explore trustworthy services in IoT. This system quickly locates services from the trustworthy nodes based on the feedbacks obtained.

Highly secure implementation done in the blockchain will help to maintain a trustful environment in the future IoT and big data era in a promising way. As an emerging technology, the combination of the Internet of Things and the blockchain has shown its great potential in solving data privacy and security problems. Thus the use of blockchain technology promises to provide an avenue of fault resistant and decentralized security solution to the big data and Internet of Things.

7 Conclusion and Future Perspectives

As we delve deeper into IoT big data, we can observe that there is a proliferation of data being generated by the things surrounding us with the increase of devices connected to the Internet. Furthermore, with the heterogeneous nature of data in big data, we must understand how the analytics and platforms are to be leveraged to correlate the data. The notion of gathering and processing the existing information remains as the biggest challenge. Besides that, if strong security measures are not applied in big data storage, it will cause some vital consequences. The speed of execution and better decision making may seem to be some other crucial steps in big data. The chapter provides an extensive review of the impact of IoT in big data emphasising on the solutions to IoT big data. We have analyzed various trust management techniques to secure big data transmission through IoT and deduced that, in order to achieve a better trade off between reliability and security, the necessity of building an efficient trust management framework will be mission critical.

As far as in-depth understanding of trust management is concerned, the current big data and IoT networks do not consider the behavioural features when deploying trust management in it. The issue behind this is that the existing trust management techniques do not scale well to establish this necessity in the IoT based big data because of the huge amount of entities as well as the limited storage and computation power. Dynamic nature of the nodes in the IoT network is also a problem to derive an acceptable level of accuracy. Therefore, a trust management mechanism that addresses these issues must be implemented to overcome the limits regarding IoT and big data. So it is necessary to understand the behaviour of devices in various situations to establish a trustworthy connection between them. In this context, it is worth considering psychology inspired techniques to provide trust management of IoT in more dynamic and practical environment. A psychological framework can effectively respond to various trust violation too in the network of IoT big data era.

References

1. Yan Z, Zhang P, Vasilakos AV (2014) A survey on trust management for internet of things. J Netw Comput Appl 42:120–134
2. Ahmed E et al (2017) The role of big data analytics in internet of things. Comput Netw 129:459–471
3. Yuan J, Li X (2018) A reliable and lightweight trust computing mechanism for IoT edge devices based on multi-source feedback information fusion. IEEE Access 6(c):23626–23638
4. Bashir AQ, Rizwan M, Gill (2016) Towards an IoT big data analytics framework : smart buildings systems. In: High performance computing and communications; IEEE 14th international conference on smart city; IEEE 2nd international conference on data science and systems (HPCC/SmartCity/DSS), pp 1325–1332
5. Ahmad A, Member S, Rathore MM, Member S, Paul A, Member S (2016) Defining human behaviors using big data analytics in social internet of things. In: Advanced information networking and applications (AINA), 2016 IEEE 30th international conference on
6. Ilapakurti A, Kedari S (2016) The role of big data in creating sense EHR, an integrated approach to create next generation mobile sensor and wearable data driven electronic health record. In: Big data computing service and applications (BigDataService), 2016 IEEE second international conference on
7. Arora D, Li KF, Loffler A (2016) Big data analytics for classification of network enabled devices. In: Advanced information networking and applications workshops (WAINA), 2016 30th international conference on, pp 708–713
8. Herwindra M, Tegar B, Rindang E, Buyung S, Wahana J, Luhung A (2016) Design and implementation of smart environment monitoring and analytics in real-time system framework based on internet of underwater things and big data. In: Electronics symposium (IES), 2016 international, pp 403–408
9. Ding Z, Gao X, Xu J, Wu H (2013) IOT-StatisticDB : a general statistical database cluster mechanism for big data analysis in the internet of things. In: Green computing and communications (GreenCom) IEEE and internet of things (iThings/CPSCom), IEEE international conference on and IEEE cyber, physical and social computing, pp 535–543
10. Lee CKM, Yeung CL, Cheng MN (2015) Research on IoT based cyber physical system for industrial big data analytics. In: Industrial engineering and engineering management (IEEM), IEEE international conference on, pp 1855–1859
11. Sezer OB, Dogdu E, Ozbayoglu M, Onal A (2016) An extended IoT framework with semantics, big data, and analytics
12. Wang H, Osen OL, Lit G, Lit W, Dai H, Zeng W (2015) Big data and industrial internet of things for the maritime industry in Northwestern Norway. In: TENCON 2015-2015 IEEE region 10 conference, pp 1–5
13. Blaze M, Feigenbaum J, Ioannidis J, Keromytis AD (1999) The role of trust management in distributed systems security In: Secure internet programming, pp 185–210

14. Ben Abderrahim O (2017) TMCoI-SIOT : a trust management system based on communities of interest for the social internet of things, pp 747–752
15. Atzori L, Iera A, Morabito G (2011) SIoT: giving a social structure to the internet of things. IEEE Commun Lett 15(11):1193–1195
16. Valarmathi AMKML (2017) Trust management in the social internet of things. Wirel Pers Commun 96(2):2681–2691
17. Hasan R, Lee AJ, Yurcik W (2005) Toward a threat model for storage systems. In: Proceedings of the 2005 ACM workshop on storage security and survivability, pp 94–102
18. Ardagna CA, Damiani E, De Capitani S, Foresti S, Samarati P (2007) Trust management. In: Security, privacy, and trust in modern data management, pp 103–117
19. Alshehri M, Hussain FK (2018) A centralized trust management mechanism for the internet of things (CTM-IoT). Int Conf Broadband Wirel Comput, Commun Appl 12(February):533–543
20. Nitti M, Girau R, Atzori L, Member S (2013) Trustworthiness management in the social internet of things. IEEE Trans Knowl Data Eng 26:1–14
21. Ben Y, Olivereau A, Zeghlache D, Laurent M (2013) Trust management system design for the internet of things : a context-aware and multi- service approach. Comput Secur 2013:1–15
22. Chen D, Chang G, Sun D, Li J, Jia J (2014) TRM-IoT : a trust management model based on fuzzy reputation for internet of things. Comput Sci Inf Syst 8(October 2011):1207–1228
23. Chen I, Guo J, Bao F (2016) Trust management for SOA-based IoT and its application to service composition. IEEE Trans Serv Comput 9:482–495
24. Chen I, Triantis KP (2018) Trust-based service management of internet of things systems and its applications
25. Jayasinghe U, Otebolaku A, Um T, Lee GM (2017) Data centric trust evaluation and prediction framework for IOT. ITU Kaleidoscope: Challenges for a Data-Driven Society (ITU K) 2017(March):1–7
26. N. Networks (2012) ITU-T
27. Bao F, Chen I (2012) Dynamic trust management for internet of things applications, pp 1–6
28. Mahalle PN, Thakre PA, Prasad NR, Prasad R (2013) A fuzzy approach to trust based access control in internet of things. In: Wireless communications, vehicular technology, information theory and aerospace \& electronic systems (VITAE), 2013 3rd international conference on, pp 1–5
29. Javanmardi S, Shojafar M, Shariatmadari S, Ahrabi SS (2014) FR TRUST: a fuzzy reputation based model for trust management in semantic P2P grids. Int J Grid Util Comput 2:1–11
30. Jayasinghe U, Lee H, Lee GM (2015) A computational model to evaluate honesty in social internet of things
31. Singh K, Verma AK (2018) A fuzzy-based trust model for flying ad hoc networks (FANETs). Int J Commun Syst 31(6):e3517
32. Bao F, Chen I (2012) Dynamic trust management for the internet of things applications, pp 1–30
33. Mahalle PN, Thakre PA, Prasad NR, Prasad R (2013) A fuzzy approach to trust based access control in internet of things. Wirel VITAE 2013:1–5
34. Bao F, Chen I, Chang M, Cho J (2012) Hierarchical trust management for wireless sensor networks and its applications to trust-based routing and intrusion detection. IEEE Trans Netw Service Manag 9(2):169–183
35. Alhadad N et al (2013) Graph-based trust model for evaluating trust using subjective logic to cite this version : HAL Id : hal-00871138 graph-based trust model for evaluating trust using subjective logic
36. Jayasinghe U, Member G, Lee GM, Member S, Um T (2018) Machine learning based trust computational model for IoT services. IEEE Trans Sustain Comput 3782(c):1–14
37. Jorge L (2015) Towards a generic trust management framework using a machine-learning-based trust model, pp 1343–1348
38. Mahmud M et al (2018) A brain-inspired trust management model to assure security in a cloud based IoT framework for neuroscience applications. Cognit Comput 10:1–10

39. Caminha J, Perkusich A, Perkusich M (2018) A smart trust management method to detect on-off attacks in the internet of things. Security Commun Netw 2018:1–10
40. Deng S, Huang L, Xu G (2016) On deep learning for trust-aware recommendations in social networks. IEEE Trans Neural Netw Learn Syst 28:1–14
41. A. Engineering (2010) TrustWalker: a RandomWalk model for combining trust-based and item-based recommendation Mohsen. In: Proceedings of the 15th ACM SIGKDD international conference on Knowledge discovery and data mining, no. 1990, pp 397–406
42. Baras JS (2009) Graph algebraic interpretation of trust establishment in autonomic networks. Prepr. Wiley J. Networks
43. Garmsiri S, Hamzeh A (2015) New graph based trust similarity measure. Ciência e Nat 37(December):339–343
44. George G, Thampi SM (2018) A graph-based security framework for securing industrial IoT networks from vulnerability exploitations. IEEE Access 6(September):43586–43601
45. Kamvar SD, Schlosser MT, Garcia-Molina H (2003) The Eigentrust algorithm for reputation management in P2P networks. In: Proc. twelfth Int. Conf. World Wide Web – WWW '03, p 640
46. Khan ZA, Ullrich J, Herrmann P (2017) A trust-based resilient routing mechanism for the internet of things
47. Nizamkari NS (2017) A graph-based trust-enhanced recommender system for service selection in IOT. In: Inventive systems and control (ICISC), 2017 international conference on, pp 1–5
48. Li L, Cai W, Liang L, Fan L (2010) Design and analysis of Bayesian game in role based trust management. In: Information theory and information security (ICITIS), 2010 IEEE international conference on, pp 394–398
49. Feng R, Che S, Wang X, Wan J (2014) An incentive mechanism based on game theory for trust management. Security Commun Netw March:2318–2325
50. Lim H, Ghinita G, Bertino E, Kantarcioglu M (2012) A game-theoretic approach for high-assurance of data trustworthiness in sensor networks
51. Fang H, Xu L, Huang X (2015) Self-adaptive trust management based on game theory in fuzzy large-scale networks. Soft Comput 21:907–921
52. Xu G, Xu C, Tian X (2012) PSO-TPS : an optimal trust path selection algorithm based on particle swarm optimization in small world network, pp 12–16
53. Huang W, Deng Z, Li R, Tang X (2013) Trust-based particle swarm optimization for grid task scheduling. Appl Mech Mater 240:1331–1335
54. Eray U, Sen S, Burak A (2015) GenTrust : a genetic trust management model for peer-to-peer systems. Appl Soft Comput 34:693–704
55. Vijayarangan S, Suresh S, Megalai J (2016) Genetic algorithm based an optimized trust based traffic analyzer for wireless sensor networks to detect malicious activities. Middle-East J Sci Res 24:41–47. https://doi.org/10.20894/IJCNES.103.005.001.010
56. Sanadhya S, Singh S (2015) Trust calculation with ant colony optimization in online social Networks. Proc Comput Sci 54:186–195
57. Sun M, Shi Z, Chen S, Zhou Z, Duan Y (2017) Energy-efficient composition of configurable internet of things services. IEEE Access 99:25609–25622
58. Gómez Mármol F, Martínez Pérez G, Gómez Skarmeta AF (2009) TACS, a trust model for P2P networks. Wirel Pers Commun 51(1):153–164
59. Bao F, Chen I, Tech V (2012) Trust management for the internet of things and its application to service composition. In: World of wireless, mobile and multimedia networks (WoWMoM), 2012 IEEE International Symposium on a World of Wireless, Mobile and Multimedia Networks (WoWMoM)
60. Wu G, Liu Z, Yao L, Xu Z, Wang W (2013) A fuzzy-based trust management in WSNs. J Internet Serv Inf … 2:124–135
61. Wang Y (2011) A probabilistic approach for maintaining trust based on evidence. J Artif Intell Res 40:221–267

62. Jøsang A (2002) The Beta reputation system. In: Proceedings of the 15th bled electronic commerce conference, pp 1–14
63. Ganeriwal S, Srivastava MB (2008) Reputation-based framework for high integrity sensor networks. Reputation-based Framew high Integr Sens networks 4:15
64. Sun Z, Zhang Z, Xiao C, Qu G (2017) D-S evidence theory based trust ant colony routing in WSN. China Commun 6:27–41
65. Che S, Feng R, Liang X, Wang X (2015) A lightweight trust management based on Bayesian and Entropy for wireless sensor networks. Scurity Commun Netw 2014:168–175
66. Qiu X, Zhang L, Wang S, Qian G (2010) A trust transitivity model based-on Dempster-Shafer theory. JNW 5(9):1025–1032
67. Feng R, Xu X, Zhou X, Wan J (2011) A trust evaluation algorithm for wireless sensor networks based on node behaviors and D-S evidence theory. Sensors 11(2):1345–1360
68. Di Pietro R, Salleras X, Signorini M, Waisbard E (2018) A blockchain-based trust system for the internet of things In: Proc. 23nd ACM Symp. Access Control Model. Technol. - SACMAT '18, pp 77–83
69. Di Pietro R, Salleras X, Signorini M, Waisbard E (2018) A blockchain-based trust system for the internet of things. Proc ACM Symp Access Control Model Technol SACMAT Part F1371:77–83
70. Hammi MT, Hammi B, Bellot P, Serhrouchni A (2018) Bubbles of trust: a decentralized Blockchain-based authentication system for IoT. Comput Secur 78:126–142
71. Zyskind G, Pentland AS (2015) Decentralizing privacy: using blockchain to protect personal data. In: 2015 IEEE security and privacy workshops, pp 180–184
72. Song JC, Demir MA, Prevost JJ, Rad P (2018) Blockchain design for trusted decentralized IoT networks In: 2018 13th Syst. Syst. Eng. Conf. SoSE 2018, pp 169–174
73. Huang Z, Su X, Zhang Y, Shi C, Zhang H, Xie L (2018) A decentralized solution for IoT data trusted exchange based-on blockchain. In: 2017 3rd IEEE Int. Conf. Comput. Commun. ICCC 2017, vol 2018–Janua, pp 1180–1184
74. Li J, Bai Y, Zaman N, Leung VCM (2017) A decentralized trustworthy context and QoS-aware service discovery framework for the internet of things. IEEE Access 5:19154–19166

A. K. Fabi is a doctoral research scholar at Indian Institute of Information Technology and Management- Kerala (IIITM-K), India. She received her M. Tech degree in Computer Science from Calicut University in 2015 and B. Tech degree in the same from the University of Kerala in 2013. Her research interests include network security, trust management in Vehicular Ad hoc Network (VANET) and Internet of Things (IoT).

Sabu M. Thampi is a Professor at the Indian Institute of Information Technology and Management, Kerala (IIITM-K), Trivandrum, India. He has completed his Ph.D in computer engineering from the National Institute of Technology, Karnataka. His research interests include network security, security informatics, bio-inspired computing, video surveillance, cloud security, secure information sharing, secure localization, and distributed computing. He has authored and edited a few books published by reputed international publishers and published papers in academic journals and international and national proceedings. He is currently serving as the Editor for 'Journal of Network and Computer Applications (JNCA)' and 'Journal of Applied Soft Computing' of Elsevier and Associate Editor for IEEE Access and International Journal of Embedded Systems, Inderscience, UK and reviewer for several reputed international journals. He is a Senior Member of IEEE and member of IEEE Communications Society, IEEE SMCS, and ACM.

Concept Drift for Big Data

Raihan Seraj and Mohiuddin Ahmed

Abstract The term "concept drift" refers to a change in statistical distribution of the data. In machine learning and predictive analysis, a fundamental assumption exits which reasons that the data is a random variable which is being generated independently from an underlying stationary distribution. In this chapter we present discussions on concept drifts that are inherent in the context big data. We discuss different forms of concept drifts that are evident in streaming data and outline different techniques for handling them. Handling concept drift is important for big data where the data flow occurs continuously causing existing learned models to lose their predictive accuracy. This chapter will serve as a reference to academicians and industry practitioners who are interested in the niche area of handling concept drift for big data applications.

Keywords Artificial intelligence · Concept drift · Big data · Cyber security · Data streams · Machine learning

1 Introduction

The advent of new technologies including IOT has enabled the availability of a vast amount of data. In fact studies show that over 2.5 quintillion bytes of data are generated each day, which reflects the online usage growth. Special field of research known as the Big Data analysis has been created that outlines techniques to systematically extract information or otherwise to deal with datasets that are very large and complex to be dealt with by traditional data processing application schemes.

R. Seraj
Department of Electrical and Computer Engineering, McGill University, Montreal, Canada
e-mail: raihan.seraj@mail.mcgill.ca

M. Ahmed (✉)
Lecturer of Computing and Security, School of Science, Academic Centre of Cyber Security Excellence (ACCSE), Edith Cowan University, Joondalup, WA, Australia
e-mail: m.ahmed.au@ieee.org

Z. Md. Fadlullah, A.-S. Khan Pathan (eds.), *Combating Security Challenges in the Age of Big Data*, Advanced Sciences and Technologies for Security Applications, https://doi.org/10.1007/978-3-030-35642-2_2

In cyber security applications the availability of these data has enabled us to use machine learning techniques and devise automated models that can detect threats. However, handling large amounts of streaming data also pose certain challenges when learning algorithms are trained on them. We present a subset of one such issue which is commonly known as "concept drift" that is evident in streaming datasets. We argue that the field of Big Data is rapidly growing and hence certain machine learning models need to be adapted accordingly. In particular, care must be taken in terms of computational cost as well as the scalability of such models.

In the area of predictive analytics and machine learning, concept drift is the statistical properties of the predicted target variable that changes over time. Due to these changes, the prediction is not always accurate. A fundamental assumption exits which reasons that the data is a random variable which is being generated independently from an underlying stationary distribution. A stationary distribution is one whose underlying statistical properties for example mean, variance etc. does not change over time. Consider a labelled dataset D, which consists of features x_i and labels y_i for each instance i. In standard classification or regression tasks, a static map f is obtained that maps the input features x_i to provide a corresponding predicted output $\hat{y}_i$. For number of instances $i \to \infty$, learning algorithms can model the underlying stationary distribution Ω of the data with bounded accuracy. The use of a static map $\widehat{y_i} = f(x_i)$ is a direct result of the stationarity of the data distribution. This means that a model learned based on historical observation pair $\langle x_i, y_i \rangle$ can be effectively used to predict future outcomes with the assumption that the relation between the input and the output remains fixed, i.e., the data are being generated from a stationary distribution. Now considering the case when the dataset D is subdivided into four parts $\{D_1, D_2, D_3, D_4\}$ where in each part, the observation pair $\langle x_i, y_i \rangle$ is being generated from an underlying stationary distribution $\{\Omega_1, \Omega_2, \Omega_3, \Omega_4\}$ respectively. Thus, the dataset D no longer consists of observations that come from a stationary distribution. Hence a static map is not sufficient to maintain a fixed input and output relation and predictive models lose their accuracy when observations come from a different stationary distribution. Also lack of sufficient examples from a particular distribution invalidates the model's power to model the distribution with a bounded error.

Concept drifts can be widely categorized into two types, namely a gradual and an abrupt drift. A pictorial representation of both the gradual and the abrupt drift is presented in Fig. 1.

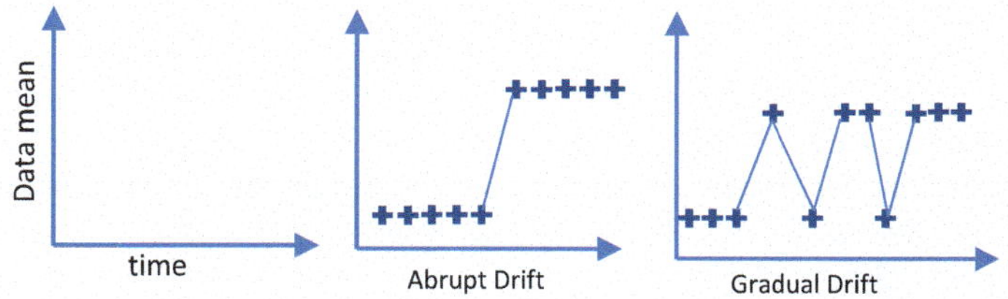

Fig. 1 Representation of concept drifts

In gradual drifts the underlying distribution of the data, slowly changes over time. A gradual shift in distribution is difficult to determine and is often a common phenomenon in time series data. In the presence of gradual drifts, examples from different distributions may occur concurrently in the dataset, hence change detection techniques are often not useful for detecting drifts under such circumstances. A more common or easily detectable from of drift includes the abrupt or sudden drift where the data distribution changes rapidly for a brief period of time and predictive models lose their accuracy at that instant. However, when abrupt drift occurs the model needs to readjust its parameters for a short time instant after which the old model can become valid again. Therefore, the detection phase for abrupt drifts is paramount for the update of the model. In the era big data and active AI solutions for cyber security, the dataset suffers from both the mixture of abrupt drift as well as gradual drift. Gradual drifts are more prominent when a model is trained on streaming data as the data flow occurs continuously. On the other hand, abrupt drift occurs for cyber security systems when the trend of the generic mode of attack is broken or when the new types of attacks originate. Many methods have been proposed for data mining in batch mode, however in the era of big data, new generation of data mining incorporates the update of the model whenever new data arrives. These techniques involve updating the model at one pass which is adaptive and thus provide a robust framework compared to its predecessors in handling concept drift for streaming data. In the subsequent sections we broadly define existing approach for handling concept drifts in the case of static datasets, we then relate the applicability of these models in the case of streaming Big data.

1.1 Chapter Roadmap

We provide a broad overview of different learning measures that are used to tackle these two natures of drifts. Most of the techniques that appear in the literature are application specific, i.e., they often include finding hand crafted features or projecting the data to different feature spaces that are resistant to drifts. With the availability of data and with the advent of deep learning, it is now possible to use state of the art deep learning techniques that are both training efficient and can tackle the problems associated with drifts. Deep neural networks can also be regarded as an automatic feature extractor which essentially alleviated the problems associated with crafting hand engineered features. The chapter gradually proceeds towards discussing simple solution techniques that appeared widely in the literature for handling concept drifts followed by a discussion on state-of-the-art deep learning solutions which appear to yield promising results. We then provide a section that gives a broad overview of adversarial drifts which is a common phenomenon in cyber security and information systems domain where we discuss different learning measures adapted in the literature in order to handle such drifts.

2 Incremental Learning

The unbounded growth of real word data and the immediate requirement of processing Big Data have posed new challenges. In streaming data, the underlying data pattern evolves over time. As a result, many dynamic learning strategies have been proposed. A popular method of training a model with streaming data involves incremental learning algorithms. The most prominent difference between incremental learning compared to other traditional learning algorithms is that the incremental learning framework does not assume the availability of sufficient training set before the learning process begins [1]. Incremental learning plays an important role for many real-world applications where the data arrives continuously over time. There are several real-world applications including user profiling, computer intrusion detection etc. where incremental learning algorithms are relevant since data arrives over time. In such a learning mechanism, the model parameters are partially updated in an incremental manner in order to accommodate the knowledge about the data based on the latest observation. In a nutshell incremental learning allows learning for streaming data where it is not possible for the model to fit all the data at once. It is a computationally efficient technique that inherently handles the problems associated with gradually. However, one of the most common tradeoffs encompassing incremental learning is the plasticity-stability dilemma. This means that there exists a tradeoff between the model's capacity to keep its knowledge that comes from older data and at the same time extend its knowledge based on latest observations. Thus, such models would seem ineffective when the data distribution is changed in a periodic manner. Due to memory constraints such models can only preserve certain knowledge built up on data coming from old distributions. Hence, they would be effective when gradual drifts are incremental, i.e., the mean distribution of the data shifts and does not return back to historical values.

Recently online incremental learning has gained much attention because they conflict the traditional assumption of complete observability of the data, additionally incremental learning framework enables updating the model parameters with few observations which makes it really suitable for online learning. Among different incremental learning algorithms, incremental support vector machine (ISVM) is the most popular one and easier to implement. The algorithm initializes limited number of support vectors (also known as the candidate support vectors) based on the available data observation. These support vectors are then updated based on new data [2]. In [3] the authors study incremental learning for robust visual tracking. Visual tracking. Drift in visual tracking occurs due to significant variation of the object's appearance or surrounding illumination. The authors presented an incremental learning approach based on principal component analysis to update the model. For a detailed analysis of different incremental learning algorithms, the reader is directed to the survey paper by [4].

3 Ensemble Learning Approach

Ensemble learning is a common machine learning paradigm where multiple learner are trained to solve the same problem of interest. Typical machine learning algorithms perform predictive analysis based on a single hypothesis which is learned from the set of training examples. In contrast, an ensemble learning approach combines a set of hypothesis to obtain a more generalised representation of the data. Ensemble learning mechanism consists of a number of learners which are usually known as base learners. The generalization ability of an ensemble is usually much higher compared to each of the base learners. Ensemble learning is appealing since it is able to boost each of these weak learners which are slightly better than random guesses and which when combined together results in a strong learner that is able to make predictions with very high accuracy [5].

Considering a classification problem, a single classifier is more robust to abrupt or sudden drifts. This is primarily due to the fact that a single classifier can quickly adapt to the new concept occurring at a particular point in time since its forgetting nature causes it to quickly respond to the new concepts. On the other hand, such classifier is not suitable when gradual drift is present in the dataset. In the presence of gradual drift an ensemble of weak classifiers is suitable. An ensemble learning mechanism in the context of a classification problem consists of many weak classifiers which increases the model diversity. A voting mechanism is used where the output label is predicted based on the decision of these multiple classifiers. It is this voting mechanism that allows handling of gradual drifts [6] since it allows the ensemble to decide the most similar concept of the data. Intuitively an appropriate number of weak learners should be chosen since increasing the number of weak learners will mean that they will not differ significantly from each other i.e., they would not be diverse [7]. Although ensemble learning algorithms are promising for handling gradual drifts, due to their inability to cope with abrupt drifts these models are often not suitable for datasets which contains both the elements of gradual and abrupt drifts. Hence much of the research has been focused to design models that can effectively handle both forms of drifts. In the era of big data, ensemble learning methods can yield significant increase in computational cost. When each individual learner is trained with large scale data, the model complexity and computational requirements grow exponentially with the amount of data. In order to tackle the computation requirements, efficient ensemble have been proposed in [8] that focuses on local learning strategy. Such strategy involves partitioning training samples into different clusters and build separate local models for each cluster. Recent works have demonstrated that a local learning strategy is better compared to that of the global learning strategy. In fact, local learning strategy enables to train ensemble for Big data thereby reducing computational cost and at the same time maintaining its performance.

4 Change Point Detection

Change point detection algorithms mainly focus on detecting abrupt drifts or specifically for anomaly detection specifically for time series datasets. Anomaly detection is often considered as a one class classification task where the main idea revolves around distinguishing between the normal and the anomalous behavior of a particular system. In the presence of concept drift, distinguishing the characteristics of legitimate situations or adversary actions become quite difficult. In the era of big data and cyber security, anomaly detection plays an important aspect in particular, such a detection scheme plays an important role in network intrusion detection [9], telecommunication fraud detection [10] or mobile masquerade detection [11]. Therefore, change point detection algorithms play an important in applications that require anomaly detection.

Change point detection algorithms can be broadly classified as an online or offline algorithm. Offline change point detection algorithms observe the entire dataset at once and then go to those instances where changes have occurred. The whole process takes data in a batch and then detect possible changes within the batch. Since for streaming big data applications, it is not possible to obtain all the data at once, one needs to follow a mechanism where algorithms can be used to detect change points whenever a dataset becomes available. Online change point detection algorithms are suitable for such cases which in contrast to offline algorithms, detect the changes right after certain number of data instances become available. Although there are subtle differences between online and off-line change point detection analysis. Online change point analysis is often used in areas such as quality control on intrusion detection and other forms of constant monitoring.

Online change point analysis	Offline change point analysis
Data arrives in single points or batches.	All the data are collected at once.
Data must be processed "on the fly" before the new data arrives.	All the data are processed at one go.
The main goal is to detect changes as soon as it occurs.	The main aim is to detect the changes as accurately as possible.
Tends to make inference about most recent changes only.	Detects all the changes that may be of interest.

In a nutshell the main challenge with online change point detection algorithm is the extent to which such algorithms can effectively and quickly detect changes for streaming data. In practice even, online change point detection algorithms are not truly real time since certain instances need to be observed in order to make decision regarding abrupt changes of the dataset.

One of the main challenges of change point detection algorithms for tackling problems associated with abrupt drift is their scalability. In the era of Big Data scalability is an important concern since the algorithms need to be adjusted to cope up with massive data in a computationally efficient manner. To determine the computational efficiency of a particular change point detection algorithm, one has to figure out whether the algorithm takes a parametric or a non-parametric form. In a parametric form, a function is learned by the model or in other words an optimal set of parameters for the function is learned by the model using a cost function. Parametric models are although simpler because gradient based approaches can be used with appropriate loss functions, such methods are often not fruitful for large datasets. Hence a popular choice in the literature has been the use of non-parametric form for very large datasets where the computational costs are significantly lower. Non-parametric form does not make an underlying assumption about the function that would define the model, instead, all the data available at the current instant needs to be stored in order to make an inference. Other important aspects for parametric change point detection algorithms is the extent to which the algorithm is sensitive to the initial parameter selection. Change point detection algorithms are often evaluated based on certain important matrices, which include accuracy, sensitivity, specificity where a drastic change in these values often indicates that a concept drift has occurred and hence the system needs to be retrained.

Change point detection algorithms contain both supervised and unsupervised method of training. In supervised models, a target or a class label is explicitly provided based on which the algorithm computes the loss of its predicted value and then tries to minimize the loss. Considering supervised schemes for change point detection of a binary or multiclass classification problem. A rolling window is used in order to slide over the available data and then compute where the data pertaining to a particular window can be considered as a possible change point [12]. While such techniques are simple during the training phase, one needs to ensure that the training phase include all possible variation of the data which is often difficult to determine. On the other hand, supervised changed point detection algorithm can also be perceived as a binary classification problem where all possible change point sequences can be considered as one class and all points with in a sequence can be considered as a second class. A common limitation of parametric change point detection algorithms is that they rely on pre-specific density models or autoregressive models whereas in real life scenario this is not often the case [13].

5 Semi-supervised and Active Learning

Task driven approaches for instance malware detection in cyber security applications, depends on large number of labelled data which contains instances and characteristics of different malware. These data are then traditionally fit into a model which is then used to predict the presence of different malwares based on the characteristics and features of the unseen data. This example portrays an application of supervised learning which is one of the most commonly used learning approaches. However, such a learning approach are used with the primary assumption that a large number of labelled examples are available. Consider a typical scenario where the nature of malware and the characteristics of these systems are evolving constantly. Malware designers are constantly trying to evade typical systems such that it becomes quite difficult to detect them in the first place. In other words, this also presents an instance of concept drift. In such a given scenario, it is not always possible to use a large number of labelled data in trying to learn a static model.

Semi-supervised learning and active learning are two mechanism that has been widely used in the literature when the inherent distribution of the data changes with time. These techniques were specially used when obtaining labelled data in the presence of concept drifts become expensive and the algorithm needs to be retrained. Semi-supervised learning algorithms are suitable for datasets containing large number of unlabeled training examples and fewer labelled training examples. This means that such an algorithm can be used along with a change point detection schemes which makes it possible for retraining the model without an explicit access to labelled data. Semi supervised learning algorithms falls in the ball park between a supervised and an unsupervised algorithm. The algorithm can be thought of as a combination of what is known as a transductive and an inductive learning. In a transductive learning scenario, the algorithms learn to assign correct labels to unlabeled data and in the inductive learning an appropriate functional mapping is learned for the input output relation. Often the use of semi supervised framework yields higher accuracy compared to solely using a supervised or an unsupervised method. In [14] a semi supervised learning approach has been used to detect and handle concept drifts in wireless signals. For a multiclass classification problem containing drifts, a simple learner for instance a random forest classifier can be used to train the model on a small undrifted labelled set. During the testing phase, the confidence measure of each of the decision stub is taken in order to assign pseudo labels to the test examples. A change point detection algorithm is used alongside which employs KL divergence as a distant metric in order to compute the distance between features. When a drift occurs, the change detection algorithm provides a feedback to the learner that a drift has occurred, the learner then incorporates the test instance with the pseudo labels in the training set and then update its parameters.

Active learning also follows similar notion as that of semi supervised learning approach. The algorithm selectively asks for labels from the expert rather than incorporating all true labels for training. Active learning measures have been extensively used as a common platform for online learning [15]. However, in the presence of drifts there are considerable differences. In case of streaming data, the decision of querying an expert for label is made at each instance. Typically, in an online setting active learning defines a fixed threshold based on uncertainly and queries the expert for labels when an instance exceeds this threshold. In the case of streaming data, due to the presence of concept drifts, this threshold property does not remain static because the threshold might be dependent based on an old concept. This makes it difficult for existing active learning measures to query the expert for labels pertaining to important regions of the instance space. Active learning with change point detection schemes can be followed in a similar fashion to that of a semi supervised learning. Considering a scenario where streaming data are available, in an active learning framework a cost is incurred upon querying the expert for a label. There is a total budget represented as which indicates the maximum total cost that the algorithm can incur by querying. As mentioned above, in the presence of concept drifts, a fixed uncertainty threshold will mean that the algorithm will exhaust its budget by having to query for a lot of labels which will eventually make the algorithm fail to learn. Therefore, when drift is present a mechanism of variable threshold needs to be incorporated in the learning framework. In [16] a variable threshold mechanism is proposed where instead of automatically labelling the instances that are less than the threshold, the algorithm labels the least certain instances with in a time interval. Once the learner becomes for certain this threshold expands automatically. In case of concept drifts when a sudden change is detected using standard change point detection algorithms or when active learning algorithm requests for a lot of labels, the threshold is contracted, and the algorithm automatically queries for the most uncertain instances. Thus, with this variable threshold mechanism, the effect of concept drifts can be handled for datasets where obtaining labels can be expensive.

6 Handling Concept Drifts with Deep Learning

Over the recent years, deep learning techniques have been quite successful in machine learning and artificial intelligence where traditional models were less competitive. Deep learning algorithms are particularly interesting since deep neural networks are considered as universal function approximations where such powerful model can extract essential nonlinear features from the dataset. In computer vision convolution neural network yield promising results with high dimensional image datasets, for time series modelling deep recurrent neural networks can be trained for prediction of sequence data. Although deep learning algorithms show promising results in high dimensional datasets such as image datasets, deep learning for information and cyber security is yet to catch up. Malware detection and network intrusion detection are the two areas where deep learning algorithms show significant improvements compared to traditional rule-based machine learning algorithms. Therefore, it is important to analyze whether rigorous feature extraction by a deep network always leads to a feature space that is invariant to drifts. While it is not straight forward to mention that deep neural network can effectively handle concept drifts since automatic feature extraction does not necessary mean that the alternate feature space would be resistant to drifts. However, several techniques have been employed in the training and the detection phase with a deep neural network for streaming data.

Algorithms where neural networks are used as a feature extractor or a parametric function approximator are often termed as “deep learning” algorithms. Handling concept drifts also remain a problem for deep learning algorithms. In a nutshell a convolution neural network (CNN) for handling concept drift. A convolution neural network is quite similar to ordinary neural network however, a CNN makes an explicit assumption regarding the nature of input data as images that allows to exploit and encode certain properties into the architecture. This also means that datasets which are not images by nature can also be used with CNN provided that they can be projected into an RGB image. Often training examples grouped together can be used and projected as an RGB image that incorporates temporal variation with in the data. More over deep architectures have a reputation of extraction essential information from an out of distribution sample. Thus, samples which are outdated due to the occurrence of concept drifts can still be used to train a deep network despite of the fact that they do not represent the current concept.

Anomaly detection for time series data has been an important task with several practical implications. Often a Recurrent Neural Network (RNN) is used for datasets containing sequential information. An RNN is a class of artificial neural network where node connections result in a directed graph along a sequence. RNN allows to capture temporal dynamic behavior for a time sequence. RNN has an internal state which are used to collect sequential input information and then process them accordingly to predict an output. Information in an RNN flows through a loop which means that in the prediction step, RNN uses the information available at the current time step and also knowledge learned from the previous inputs in order to make a decision. The output of the RNN is then again copied and looped into the network to make prediction at the next stage. Traditional anomaly detection techniques used to learn a model for normal behavior of the time series, which are then used to detect any anomaly for unseen data. However, it is quite important that anomaly detection algorithms adapt when there is a change in concept that results in the underlying distribution being non-stationary. Much of the focus on anomaly detection using RNN has been in the offline phase for time series data, where RNN is trained on historical events with the fundamental assumption that the training and the test data come from the same distribution. As concept drift becomes a crucial problem for generic time series data resulting in non-stationarity training of an RNN needs to be adapted. In [17] the authors propose an incremental training approach for RNNs which essentially makes the training online. RNNs are trained on incoming data streams for a time series and multi-step predictions are made based on historical data. These prediction errors are then used to fine tune the model and tackle the problems associated with concept drifts. In addition to it, a local normalization technique is used over a window that also ease the effect of non-stationarity of the data. Since training RNNs are quite difficult due to the common problem known as the vanishing gradient problem, in order to make training stable, RNN with Long Short-Term Memory (LSTM) cell or Gated Recurrent Unit (GRU) cell are used. A comprehensive detail on LSTM and GRU can be obtained in [18, 19] which is beyond the scope of this chapter.

A more recent analysis for concept drifts with deep neural networks lies in the network architecture itself. Spiking neural networks, first introduced in [20] open new doors towards finding reliable solutions for concept drifts. The model has been inspired from the neuroscience literature with the need to better understand the information processing schemes in a mammal brain. It does not suffer from the common draw back faced by most traditional algorithms, i.e., retraining when a change in concept is detected. Spiking Neural Networks (SNN) have been one of the most potential models that can actively represent the behavior and the learning potential of the human brain. SNN makes a learning rule based on spike representation of the information. This spiking time learning mechanism essentially allows the network to capture temporal association between a large number of variables in a streaming data. A more successful spiking neural network has been the Evolving Spiking Neural Network, where the number of spiking neurons grows with time in an incremental manner so as to capture temporal patterns from the data. In [21] the authors present an online training phase and an enhanced model of Evolving Spiking Neural Networks that introduces a sliding window and deals with the limitation of the size of the neuron repository. A hybrid model is proposed along with a drift detection mechanism to be used with the Evolving -SNN so that drift can be detected and corrected for online. While this chapter only provides a high-level intuition of different mechanisms that are being used for concept drift, it does not provide a rigorous analysis of each of the network architecture. Hence readers particularly attracted to this model are referred to [22] in order to obtain a complete architectural review for SNN. Although the concept of SNN for handling drift remains new, their application and modelling framework yields promising outcomes opening the doors towards new research directions.

7 Beyond Gradual and Abrupt Concept Drifts

This section of the chapter focuses on a special type of drift, that is present in cyber and information security domain. In particular we mainly focus on a special class of drift which is known as an Adversarial drift. Adversarial drifts are a special class of drift that needs specific attention since off the shelf methods for handling concept drifts are not effective for adversarial drifts as they are designed from an adversarial agnostic perspective. In an adversarial drift, an active adversary tries to evade the actively deployed learning model. Thus, when model parameters are updated with the primary assumption that drift is a benign change, they fail to work in the presence of an active adversary. The main characteristics of adversarial drifts are a) the drift is a result of changes to malicious class only b) drift is a function of the classification model, since the adversary learns about the model in order to gain information before trying to evade it c) The drift is always targeted to hinder the performance of the existing learning algorithm (i.e., the drift leads to a drop in performance such as accuracy, predictive confidence measure of the deployed model) [23].

Adversarial drifts play a significant role in hindering the performance of a network intrusion detection systems where adversaries change the signature of the malware in order to evade the traditional intrusion detection systems.

Deep learning-based system with the concept of self-taught learning has proved to be quite successful for detecting unknown network intrusions in [24]. Deep learning techniques have proved to have better generalization capability in the presence of adversarial drifts, compared to traditional machine learning algorithms. In [23] the authors presented the predict detect framework that handles adversarial drifts for streaming data. The predict detect system signals the presence of adversarial drifts with high reliability along with a lifelong learning framework. The robustness of a model to adversarial attacks depends on the ease at which the drifts can evade such models. Therefore, it is necessary to obtain maximum feature information for such systems. The intuition behind extracting maximum information from the features comes from the fact that since the model now has more information, it becomes difficult for the adversary since now it has to have properties where it can cause a breach to the system and at the same time it needs to maintain the characteristics of the benign data so as to fool the model from detecting it in the first place. As was the case mentioned in the previous section where important feature information is paramount for extracting knowledge about the data that would remain stationary over time, in the case of detecting and maintaining the model performance for adversarial drifts an equal importance for extracting maximum feature information is also justified.

In the case of streaming data, this notion of gaining maximum information from the features become difficult since one has to now take the time delay between detecting a drift and correcting the model for it into account. In fact, it is important that adversarial drifts are detectable quickly and the models are updated promptly as well in order to make them useable over time. Therefore, complex learning strategies need to be adapted in order to take these situations under consideration. In the case of streaming data, obtaining more information about the features of the training data, can often lead to a phenomenon known as the leakage of information to the adversary. This means that the adversary now has important information and can channel the attacks accordingly. Thus, it is very important that during the testing time of a model for detecting adversaries, the test is done on samples from the dataset that is completely separate from that of the benign training set [23]. The predict detect approach proposed in [23] uses two orthogonal classifiers which are trained on disjoint subset of features of the training data. The first subset of features is used to train a model which is known as the *Prediction* model that perform predictions on the input samples. Over time after deployment, this model is expected to get attacked due to the nature of the adversarial environment. It is then when the second model which is known as the *Detection* model that had been trained on second subset of features become useful as they still remain "hidden" to the adversary since it is not used to for the prediction in the first place. Finally, the *Detection* model declares an adversary based on the disagreement with the *Prediction* model.

Other works on detecting adversarial drifts in security games include developing meta learner that will weight its actions against an adaptive adversary evolving its tactics in response to the learner's predictions, [25]. Ensemble in adversarial classification for spam has also been studied in [26] where the authors present a method based on ensemble of classifiers that can detect degradation of its performance and hence can retrain themselves without manual intervention. In [27] the authors present a statistical test to distinguish the adversarial examples from the model's training data, the statistical test has an important property of being model agnostic, at the same time the authors present an integrated outlier detection scheme that adds an additional class to the model's output and train the model to recognize adversarial examples as part of this new class.

This section mainly provides a broad overview of how adversarial drifts occur and include some measures of how they can be mitigated. While the literature concerning adversarial drifts are more diverse, comprehending all the solution techniques widely adapted in the literature is beyond the scope of this chapter, for more details the reader is redirected to a comprehensive study presented in [28].

8 Conclusions

We live in a very interesting time where the technological advancement created some challenges such as cyber-attacks, big data, predictive analysis etc. We have discussed an important issue called concept drift in this chapter and tried to explore different solutions relevant to the issue of concept drift. In recent years, deep learning is considered to be a great solution for any modelling tasks; therefore, we also explored the deep learning for the concept drift issue for solving big data problems. The contribution in this chapter will help the graduate students, researchers in the area of artificial intelligence to explore more about the issue of concept drift and propose more effective solutions.

References

1. Zang W, Zhang P, Zhou C, Guo L (2015) Comparative study between incremental and ensemble learning on data streams: case study. J Big Data 1:5
2. Cauwenberghs G, Poggio T (2001) Incremental and decremental support vector machine learning. Johns Hopkins University, Baltimore
3. Ross DA et al (2008) Incremental learning for robust visual tracking. Int J Computer Vision 77(1–3):125–141
4. Losing V, Wersing BHH (2018) Incremental on-line learning: a review and comparison of state of the art algorithms. Neurocomputing 275:1261–1274
5. Oza NC (2001) Online ensemble learning. University of California, Berkeley
6. Liao J-W, Dai B-R (2014) An ensemble learning approach for concept drift. In: Information science and applications (ICISA), 2014 international conference on. IEEE
7. Gomes HM (2017) A survey on ensemble learning for data stream classification. ACM Computing Surveys (CSUR) 50(2):23
8. Yoo PD, Ho YS, Zhou BB, Zomaya AY (2008) SiteSeek: post-translational modification analysis using adaptive locality-effective kernel methods and new profiles. BMC Bioinformatics 9:272
9. Lee W, Stolfo S, Mok K (2000) Adaptive intrusion detection: a data mining approach. Artif Intell Rev 14(6):533–567
10. Hilas CS (2009) Designing an expert system for fraud detection in private telecommunications networks. Expert Syst Appl 36(9):11559–11569
11. Mazhelis O, Puuronen S (2007) Comparing classifier combining techniques for mobile-masquerader detection. In: The second international conference on availability, reliability and security
12. Aminikhanghahi S, Cook DJ (2017) A survey of methods for time series change point detection. Knowl Inf Syst 51(2):339–367

13. Kawahara Y (2009) Change-point detection in time-series data by direct density-ratio estimation. In: Proceedings of the 2009 SIAM international conference on data mining. Society for Industrial and Applied Mathematics
14. Ghourchian N, Allegue-Martinez M, Precup D (2017) Real-time indoor localization in smart homes using semi-supervised learning. In: AAAI
15. Cohn D, Atlas L, Ladner R (1994) Improving generalization with active learning. Mach Learn 15(2):201–221
16. Zliobaite I, Bifet A, Holmes G, Pfahringer B (2011) MOA concept drift active learning strategies for streaming data. In: Proceedings of the second workshop on applications of pattern analysis
17. Saurav S (2018) Online anomaly detection with concept drift adaptation using recurrent neural networks. In: Proceedings of the ACM India joint international conference on data science and management of data, ACM
18. Hochreiter S, Schmidhuber J (1997) Long short-term memory. Neural Comput 9(8):1735–1780
19. Cho K (2014) Learning phrase representations using RNN encoder-decoder for statistical machine translation, In: arXiv preprint arXiv:1406.1078
20. Gerstner W, Kistler WM (2002) Spiking neuron models: Single neurons, populations, plasticity. Cambridge University Press, Cambridge
21. Lobo JL et al (2018) Evolving spiking neural networks for online learning over drifting data streams. Neural Netw 108:1–19
22. Budiman A, Fanany MI, Basaruddin C (2016) Adaptive convolutional ELM for concept drift handling in online stream data. In: arXiv preprint arXiv:1610.02348
23. Sethi TS, Kantardzic M (2018) Handling adversarial concept drift in streaming data. Expert Syst Appl 97:18–40
24. Niyaz Q, Sun W, Javaid AY, Alam M (2016) A deep learning approach for network intrusion detection system. In: Proceedings of the 9th EAI International Conference on Bio-inspired Information and Communications Technologies (formerly BIONETICS) ICST (Institute for Computer Sciences, Social-Informatics and Telecommunications Engineering)
25. Abramson M (2015) Oward adversarial online learning and the science of deceptive machines. In: AAAI fall symposium series
26. Chinavle D et al (2009), Ensembles in adversarial classification for spam. In: Proceedings of the 18th ACM conference on Information and knowledge management, ACM
27. Grosse K et al (2017) On the (statistical) detection of adversarial examples. In: arXiv preprint arXiv
28. Kantchelian A et al (2013) Approaches to adversarial drift. In: Proceedings of the 2013 ACM workshop on artificial intelligence and security, ACM

Raihan Seraj is currently pursuing his PhD in the department of Electrical and Computer Engineering at McGill University. He attained his Master's degree from the same university where he worked on learning algorithms in the presence of partial observability and concept drifts. His research interest include reinforcement learning for multiagent systems, deep learning for real time indoor localization with wifi signals. He is actively involved in the fields of machine learning and artificial intelligence.

Mohiuddin Ahmed attained his PhD from UNSW Australia and currently working as Lecturer in the Academic Centre for Cyber Security Excellence at Edith Cowan University, Australia. His research interests include Big Data Mining, Machine Learning and Network Security. He is working to develop efficient and accurate Anomaly Detection techniques for network traffic analysis to handle the emerging Big Data problems. He has made practical and theoretical contribution for data summarization for network traffic analysis. His research also has a high impact on critical infrastructure protection (SCADA systems, Smart Grid), information security against DoS attacks and complicated health data (heart disease, nutrition) analysis. He has published a number of journals and conferences papers in reputed venues of computer science. Mohiuddin holds a Bachelor of Science Degree in Computer Science and Information Technology with High Distinction from Islamic University of Technology, OIC.

Classification of Outlier's Detection Methods Based on Quantitative or Semantic Learning

Rasha Kashef, Michael Gencarelli, and Ahmed Ibrahim

Abstract The problem of outliers (Anomalies) detection has been generally presented as a single-minded problem, in which outliers are defined as objects that do not conform to a given definition. In this chapter, we propose a novel taxonomy that groups the methods into two categories: (1) quantitative outlier detection and (2) semantic outlier detection. For quantitative outliers, outliers are defined based on a calculated outlier score. For semantic outliers, there is a conceptual meaning behind the outlier based on the context of the dataset, shifting the focus to finding the anomalous class of data. We also discuss the use of the proposed definition of semantic learning in detecting credit card frauds.

CCS CONCEPTS
Computing methodologies → Anomaly detection

Keywords Outliers detection · Quantitative and semantic learning

1 Introduction

Outlier detection has traditionally been viewed as a one-dimensional problem, and improving the accuracy of one technique leads to improvements in outlier detection as a whole. The purpose of this paper is to provide a summary of current research

R. Kashef (✉)
Department of Electrical, Computer, and Biomedical Engineering, Ryerson University, Toronto, ON, Canada
e-mail: rkashef@ryerson.ca; https://www.ee.ryerson.ca/people/Kashef.html

M. Gencarelli
IVEY Business School, London, ON, Canada
e-mail: mgencarelli.msc2018@ivey.ca

A. Ibrahim
Computer Science Department, Western University, London, ON, Canada
e-mail: aibrah64@uwo.ca

Z. Md. Fadlullah, A.-S. Khan Pathan (eds.), *Combating Security Challenges in the Age of Big Data*, Advanced Sciences and Technologies for Security Applications, https://doi.org/10.1007/978-3-030-35642-2_3

on outlier detection. We fundamentally propose two types of outliers that can be detected: (1) quantitative outliers, and (2) semantic outliers.

Quantitative outliers are defined as those that deviate from other data points based on a quantitative metric. These outliers are independent of the context of the dataset, meaning that any detected outlier is based entirely on some numeric values calculated from the data. Semantic outliers place a metaphysical definition of what it means to be an outlier. Rather than quantifying an outlier, we can conceptualize that an outlier will belong to a specific class in the data, and train an algorithm to identify observations that belong to that class. Each outlier detection method defines one type of outlier.

We define a novel taxonomy that groups outlier detection methods as either quantitative or semantic detection methods. The rest of this chapter is organized as follows: in Sect. 2, our taxonomy of outlier's detection algorithms is presented. Quantitative outlier's detection is defined and discussed in Sect. 3. Semantic outlier's detection is presented in Sect. 4. Section 5 provides insights into the application of semantic learning in credit card fraud detection. Finally, we conclude the paper in Sect. 6.

2 Taxonomy of Outlier Detection

Traditional Outlier is defined as a data point that deviates from other data points as to arouse suspicion that it was generated by another mechanism [8, 10]. The definition implies that all datasets will mainly have "normal" data and only a selected few objects will be outliers. This definition is impractical on its own for detecting outliers because it does not offer criteria to identify "normal" versus "abnormal" data. The problem is that "normal" is a relative concept. By examining the outlier detection techniques proposed in the literature, each technique makes an assumption about what "normal" data should be. We have found that these assumptions can be grouped into two distinct categories that represent the type of outlier being detected. The first option, quantitative learning, is to look at a dataset as a standalone set of observations and assign each data point a quantitative outlier score based on an evaluation metric. Normal data points will have an outlier score that falls in an expected range, while outliers would fall outside of the range. This approach can be applied universally to any dataset across many applications, and the underlying calculations with the data remain the same. The second approach is to think of normal data within an application-specific context. We formally define these two types of outliers as an extension of the traditional definition.

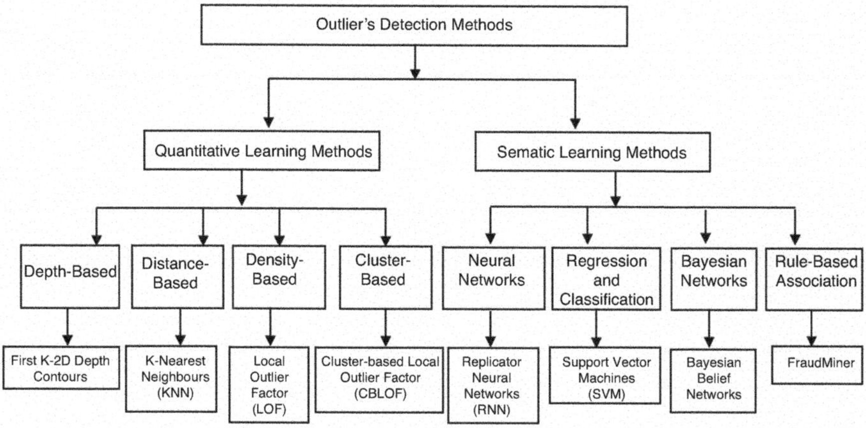

Fig. 1 Taxonomy of outliers detection methods

Definition: *Quantitative Outlier*

A Quantitative Outlier is a traditional Outlier such that for any dataset D, an object P will be an outlier in D if it is different from all other data points based on a quantitative evaluation metric M.

Definition: *Semantic Outlier*

A Semantic Outlier is a traditional Outlier such that for an application space A, an object P will be an outlier in A if it belongs to the anomalous class C.

In Fig. 1, we define a novel taxonomy for grouping outlier detection techniques as either quantitative or semantic.

3 Quantitative Outlier Detection

In this section, we discuss those outlier detection techniques that identify quantitative outliers. The common characteristic of these algorithms is that they are independent from the context of the dataset. Regardless of which application the data is from, outliers are computed using a predefined quantitative metric M. This results in an outlier score that measures the degree of outlierness for each data point. Each algorithm applies a unique computational strategy to calculate the outlier score. Classification is performed by comparing the outlier score to a user-defined cutoff threshold, or by taking the top n objects with the largest score. The limitation with

Table 1 Properties of a quantitative outlier

Global	Local
Distance-based	Density-based
Depth-based	Cluster-based

the quantitative approach is that the outlier score is only relevant for the current data. If more data is provided afterward, the algorithm would need to be recomputed to obtain the new outlier scores for every data point. This means that the results from the algorithm are static and cannot be extrapolated to future data. Each quantitative algorithm detects a unique subset of outliers based on their properties. First, the algorithm must consider the relative positioning of a data point against the others. If the focus is on detecting global outliers, then the algorithm will detect those points that deviate the most from every other point. This is not sufficient in some applications. Instead, local outliers can be identified by evaluating the degree of outlierness against the data within the local neighborhood. As a result, a point may not be considered a global outlier, but within the same data, it could be considered a local outlier. The second property is the number of points that are clustered together to form an outlier. The traditional notion of an outlier is that it represents a single object, but in some cases, it may be more appropriate to label groups of outliers when a similar anomalous event occurs repeatedly. We can divide the quantitative algorithms into subcategories based on the properties of the quantitative outlier as shown in Table 1.

3.1 Depth-Based Algorithms

The depth-based approach considers the points at the outermost edge of the data to be the most likely to be outliers. The algorithm assigns a depth score that increases towards the innermost data objects and then identifies groups of global outliers based on the data with the lowest depth score.

3.1.1 Convex Hull – First K-2D Depth Contours

The First K-2D Depth Contours algorithm uses convex hulls (As shown in Fig. 2) [11] to establish the depth of all points in a dataset. Each hull is constructed iteratively by finding the smallest bounding polygon such that every point exists either on the boundary of one of the hulls or within its interior. The outer most layer has a depth of 1, and the depth increases for each of the interior hulls. If a data point has a higher depth, it is considered less likely to be an outlier. The convex hull algorithm is efficient for contouring up to a three-dimensional space, but in higher dimensions, the computational complexity increases drastically [11].

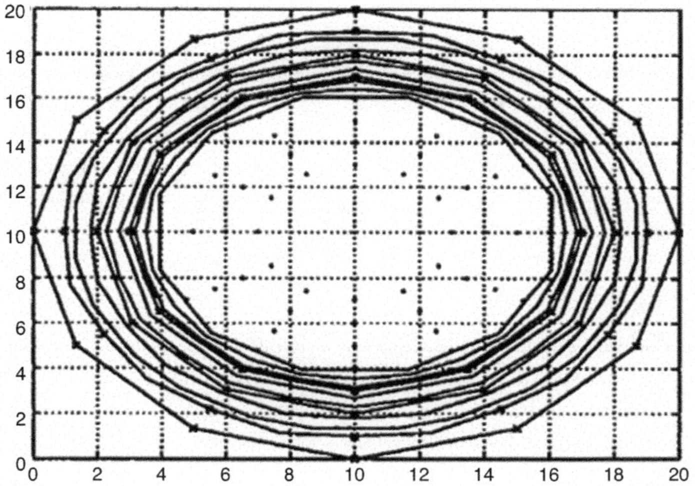

Fig. 2 Depth contours [11]

3.2 *Distance-Based Algorithms*

A distance-based outlier is defined as an object with at least a fraction p of its neighboring objects lying greater than a distance D [12]. The differentiating characteristic of each algorithm is the measure that is used to assess proximity, which is used to calculate a score of outlierness for each point. The top *n* points with the highest score are then considered outliers, or a cutoff threshold is used.

3.2.1 Spatial Proximity

Spatial proximity involves a distance calculation between a single point and the ones surrounding it. The algorithm defines two parameters: ε is the radius of a bounding circle that encompasses all data points, and π is a percentage threshold. A point P will be considered an outlier if at most π percentage of points have a distance to P that is less than ε. P is, therefore, an outlier when the majority of points are further away than the distance threshold. Spatial proximity calculations are straightforward and applicable to a wide range of applications, but they lack efficiency because the distance needs to be calculated for every data point.

3.2.2 K-Nearest Neighbours (KNN)

The KNN approach can be applied in a variety of ways to detect outliers. The simplest approach is to calculate the total distance from the object to its 1-NN, 2-NN, …, K-NN and use this to calculate an outlier score for each data point. Thus, either

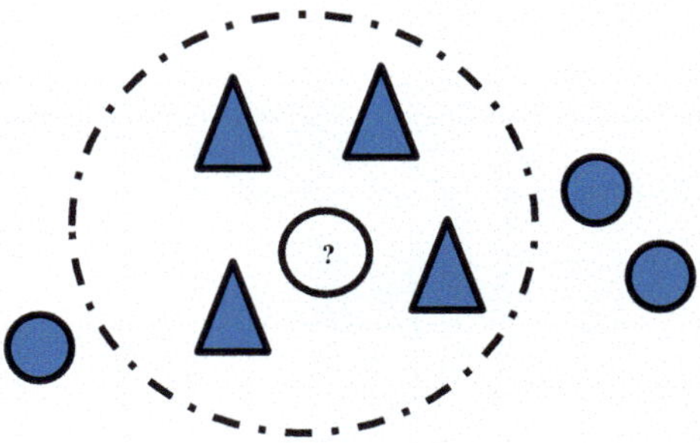

Fig. 3 Majority vote in 5-NN classifier

the top *n* scores can be taken as outliers, or a distance threshold can be put in place. In comparison with spatial proximity, KNN lacks efficiency because it still requires distance calculations to be made to each data point. Figure 3 shows a classification of data point as an outlier by majority votes from its 5 nearest neighbors.

3.2.3 Orca – Improved KNN Approach

Orca is an improvement over the simple KNN approach that was designed to improve the efficiency of distance-based algorithms. The output from this approach is a list of the top *t* data points that have been classified as outliers [3]. The algorithm works by keeping track of a cut-off threshold C, which represents the outlier score for the t-th largest outlier. t is a user-defined parameter for the total number of outliers for the algorithm to identify, and the threshold C contains the lowest outlier score from the list of current outliers. When the algorithm begins, t data points are randomly selected to populate the list of outliers and their outlier scores are calculated based on the distance to their k nearest neighbors. The cut-off threshold C is updated with the outlier score of the t-th largest outlier. The algorithm then begins to process each of the other points. It does this by calculating the distance to each of its nearest neighbors, and then dynamically comparing the point's outlier score to the cut-off threshold C. If at any time the data point's current outlier score drops below the threshold, the data point is pruned because it can no longer be considered an outlier. The Orca method improves the efficiency of distance-based algorithms by dynamically eliminating points that are unlikely to be outliers. This reduces the number of distance calculations that need to be processed [3].

3.2.4 In-Degree Method

Graph theory can be applied to outlier detection in a variation of the KNN approach. We saw that in a traditional KNN approach, an outlier score is calculated based on the distance from a point to its nearest neighbors. A variation of this approach replaces the scoring approach with a directed graph. For a data point P, each of its k-nearest neighbors will have a directed edge drawn from P to itself. The user will then define a threshold t for the in-degree of each data point, and if a point doesn't meet this threshold, it will be considered an outlier. The assumption, in this case, is that an outlier is any point that is not the nearest neighbor of at least t other points.

3.3 Density-Based Algorithms

Density-based algorithms are used to find densely-populated neighborhoods of data. When a data point is located in close proximity to many other points, these data points form a cluster and are less likely to be outliers. In contrast, areas with a sparse distribution of points are more likely to be outliers. Density-based algorithms calculate an outlier score based on the neighborhood that it is found. For this reason, they can be used to find single, local outliers.

3.3.1 Local Outlier Factor (LOF)

The local density of a data point is calculated by determining the reachability of an object p with respect to another object o. This is calculated by first determining the k-distance of an object o, defined as the distance that contains all of its k-nearest neighbors. If an object is within the k-distance, its actual distance is replaced by the k-distance, and this becomes its reachability. If the object is outside of the k-distance, its actual distance is used. Based on this concept of reachability, if an object p is outside the k-distance of object o, it will have a sparse local density with respect to object o. The LOF is computed by comparing a point's local density with the local densities of its neighbors. If an object is found within a densely populated cluster, its local density will be comparable to that of its neighbors, and therefore the LOF ratio will be close to 1. For objects that are located slighted outside of a densely populated cluster, their local density will be slightly lower, therefore leading to a higher LOF. Any object that is significantly outside of a cluster will have a significantly higher LOF. The LOF, therefore, becomes a scoring mechanism to determine the degree of outlierness of an object with respect to other objects within its neighborhood as shown in Fig. 4. Depending on the application, an upper bound can be placed on the LOF to determine how sparse of a local density the user is willing to tolerate before considering the data point an outlier [4].

3.4 Cluster-Based Algorithms

Cluster-based algorithms are similar to density based algorithms, but they can identify small clusters of outliers rather than point outliers. The algorithms assign an outlier score based on the cluster that they belong to, rather than at an individual level.

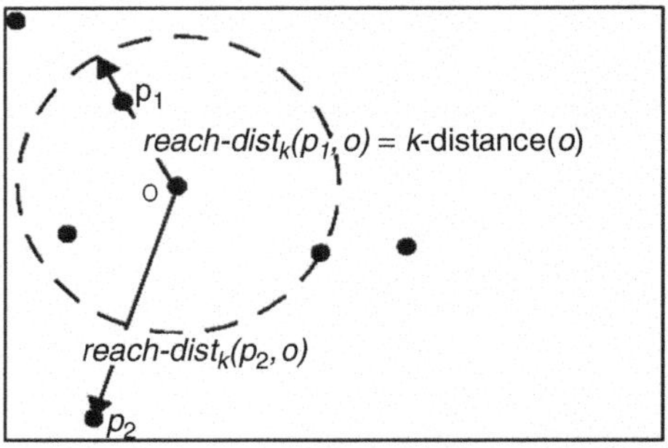

Fig. 4 Reachability-distance in LOF method [4]

FindCBLOF Algorithm
Step1: Partition the dataset into *k* clusters using any clustering technique A_i.
Step2: Assign an outlier factor to each object is, namely, *CBLOF*, which is a measure of both the size of the cluster the object belongs to and the distance between the object and its closest cluster (if the object lies in a *small* cluster)
Step3: Rank objects based on their CBLOF and return %TopRaio objects with the highest CBLOF as outliers

Fig. 5 FindCBLOF

3.4.1 Cluster-Based Local Outlier Factor (CBLOF)

For outlier detection, CBLOF identifies outliers based on the assumption that normal data points should fall within a large cluster. It is less probable for an outlier to reside in any of the large clusters. In order to formally define the difference between a large cluster and a small cluster, the algorithm checks two conditions: (1) if the number of points in a cluster Ci is above a numeric parameter, and (2) if the ratio between cluster Ci and cluster Ci+1 is greater or equal to a second numeric parameter. If one of these conditions holds true, then cluster Ci becomes the boundary between large and small clusters. The algorithm begins by using a clustering approach to generate the clusters, and then cluster-based outliers are identified by using the cluster-based local outlier factor (CBLOF). The CBLOF is calculated by comparing the distance between each point and that of the nearest large cluster. By doing this, points that have been captured within a large cluster should have a low CBLOF. However, points that are within a small cluster should all have a large CBLOF. These points are more likely to be classified as cluster-based outliers [14]. The FindCBLOF algorithm is shown in Fig. 5.

3.4.2 LDBSCAN

The LDBSCAN algorithm computes density-based clusters using the LOF measure discussed previously. To identify cluster-based outliers, only clusters with the smallest number of data points are considered based on an upper-bound threshold. The cluster-based outlier factor (CBOF) is computed for each cluster by multiplying the number of objects in a small cluster C by the product of the distance between C and its nearest cluster and the average local reachability of the nearest cluster. This means that the CBOF will increase based on three factors: (1) the more objects contained in C; (2) the further away it is from its nearest cluster; and (3) the greater the density of the nearest cluster [7].

4 Semantic Outlier Detection

In this section, we introduce semantic outlier detection to identify outliers within an application-specific context. Semantic outliers have a pre-established meaning behind them that can be labeled as class C. The purpose of these algorithms is to learn the behavior of the normal and anomalous classes so that when presented with new data, the model should be able to accurately classify the data between the two classes. The algorithms first train a model using existing data so that it can learn to recognize the patterns of normal attributes. When presented with new data after training, the model should be able to predict which data points are normal and which are anomalous. Each semantic detection method can be used in one-class and multi-class training. With one-class techniques, the model is only being trained on the normal behavior. If the model then encounters an object that does not fit the learned pattern for normal data, it will classify it as an outlier. This is a useful method when the characteristics of the known anomalous class C are not always known. In contrast, with multi-class techniques, the algorithm is learning to differentiate between both the normal and anomalous behavior. The outlier detection problem, therefore, reduces to a classification problem. If supervised learning is used, we can compare the classification from the model with the actual data and determine the accuracy of the model's predictions.

4.1 Replicator Neural Networks

Replicator Neural Networks (RNN) are feed-forward, backward propagation perceptron with the purpose of reproducing the same input vector at its output. During the learning phase, the training data passes through the network, and the reconstruction error is calculated. The error term is propagated backward through the network, and the weights are adjusted with the objective of minimizing the error for the next iteration. The network learns to reproduce common patterns in the data with high precision, but it will have difficulty reproducing abnormal patterns that are expected to be outliers. The RNN model was proposed by [6] to detect semantic outliers in the one-class setting. The model was trained using only normal data points with the intent to learn the normal pattern of behavior and classify anything that deviates from the normal as an outlier. The largest reconstruction error, therefore, becomes a threshold for outlierness. During testing, if a new data point results in a higher error than this threshold, then it is considered an outlier. The RNN network was evaluated

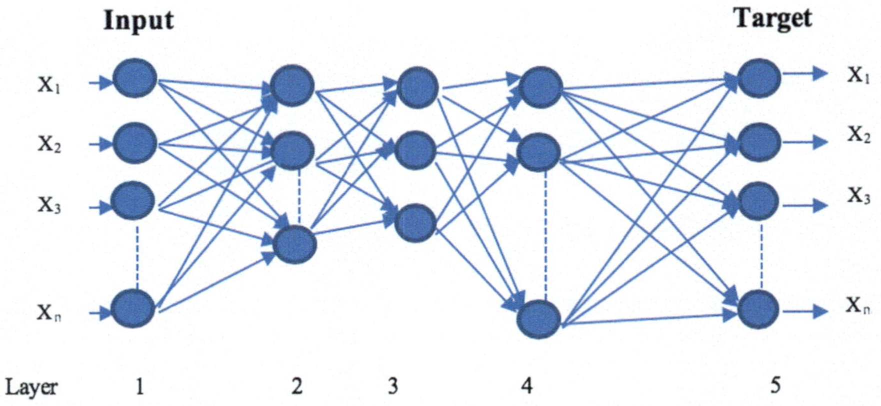

Fig. 6 Replicator Neural Networks (RNN)

on a public breast cancer dataset where the data points were labeled either as malignant or benign, with malignant representing an outlier. The advantage of the one-class RNN is that it was more accurate in identifying the malignant cases than its multi-class counterpart. The supervised RNN was able to detect all malignant cases in the top 48 records while the unsupervised RNN could only detect 89.74% [6]. An example of a RNN with three hidden layers is shown in Fig. 6.

4.2 Support Vector Machines

Support Vector Machines (SVMs) were first proposed by Schölkopf as a regression and classification algorithm with a strong generalization capability [17]. In outlier detection, the SVM has been primarily proposed as a one-class outlier detector. The one-class SVM is used to create a geometric boundary between the normal data points and the outliers. An optimization problem is solved to maximize the margin between the two classes, improving the generalization ability of the model. After training, any object that falls within the boundary are classified as normal while those that fall outside the boundary are classified as outliers. The algorithm begins by mapping the data to a higher-dimensional feature space using a non-linear transformation based on a kernel function, defined as $\varphi: X \rightarrow F$. The mapping is required to allow for the separation of data in a feature space which would typically be inseparable in the input space. The boundary around the data is constructed by solving for the parameters that will form the geometric shape – a hyperplane and the hypersphere are typically used in outlier detection, although a quarter-sphere has also been proposed by [13]. The parameters for the optimization problem depend on the shape of the boundary. If a hyperplane is used, the parameters to solve for are the weight vector w which represents the vector normal to the hyperplane, and a bias parameter r. If a hypersphere is used, the parameters are the centroid of the sphere

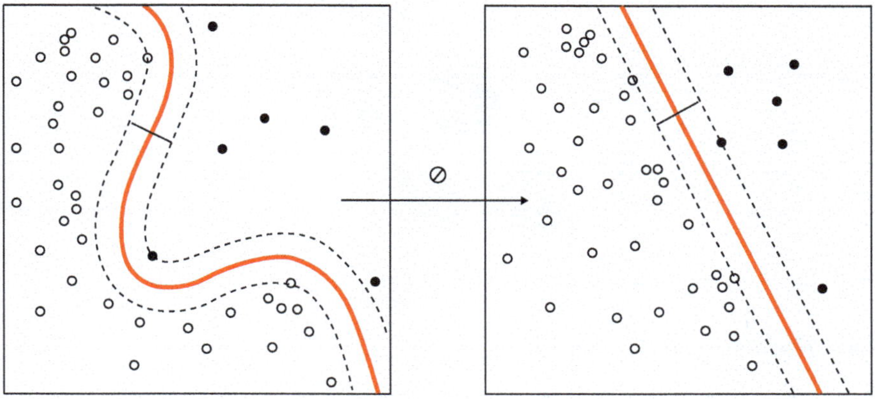

Fig. 7 One-class SVM (https://en.wikipedia.org/wiki/Support-vector_machine)

and its radius [19]. The generalization ability of the model is characterized by the values of a regularization parameter ν and a slack parameter ξ. The slack parameter represents the cost of misclassifying an object, then the sum of the slack parameter for all data points represents the total cost of misclassifying an object. The regularization parameter determines the fraction of objects that are allowed to be misclassified. If the total cost of misclassifying an object is low, then more data will be classified as outliers – leading to a higher detection rate but also a higher false-positive rate [19] Fig. 7.

4.3 *Bayesian Belief Networks*

Bayesian Belief Networks are used to construct a probability density model for multivariate data [15]. The assumption is that the data contains attributes that are conditionally dependent on each other, and these conditional dependencies determine the likelihood of observing an outcome from the attributes. A Bayesian network is a directed, acyclic graph along with an associated set of probability tables for each node n $\in$N [5] as shown in Fig. 8.

The nodes represent a set of random variables with mutually exclusive states. Given two random variables A and B, if a directed edge exists from nA to nB then the state of A is conditionally dependent on the state of B, and the likelihood of observing A is the posterior probability after knowing B. Bayesian networks can be used for multi-class outlier detection by determining the likelihood that an observation belongs to an anomalous class. For each observation, the attribute values are the inputs into the Bayesian network, and the output is the probability that the observation belongs to a specific class. If the model predicts that there is a greater probability for observing an anomalous class for A, the record will be classified as an outlier.

P(A=1)=0.6	P(B=0)=0.02

A	B	P(C=1\A,B)
1	1	0.6
1	0	0.08
0	1	0.45
0	0	0.85

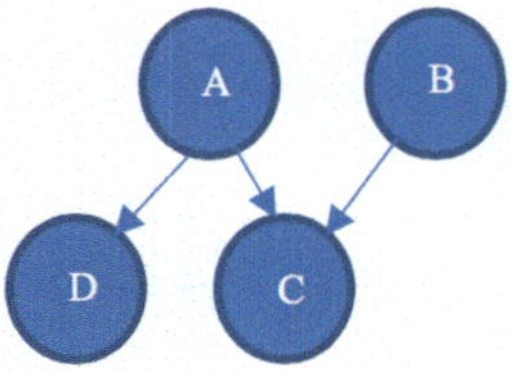

A	B	P(D=1\A)
1	1	0.55
1	0	0.75

Fig. 8 Bayesian network of four variables with 2 states

4.4 Rule-Based Approach

The rule-based approach identifies common patterns in the dataset features using frequent itemset mining [1]. An itemset is first defined by its length, representing the number of attributes and their values that are included in the itemset. In general, larger itemsets are considered to be stronger rules for classification because it incorporates more features. The defining characteristic of the itemset is its support value, defined as the percentage of transactions that coincide with the rule. For outlier detection, frequent itemsets can be found for both normal and anomalous data, and a new data point will be classified based on the rule that it matches. FraudMiner is a rule-based approach for detecting frauds in credit cards [18].

5 Credit Card Fraud Detection

Credit card fraud is a specific application of outlier detection where a fraud transaction is labeled as an outlier [2]. This special case follows the semantic type of outlier detection, where a fraud transaction is labeled as a semantic outlier. In this section, we examine the use of semantic in detecting credit card fraud. The RNN model was first used to detect credit card fraud in the CardWatch application (Aleskerov, Freisleben, and Rao). CardWatch was used to learn the spending profile of each customer, and therefore an independent network is needed for each customer. An artificial dataset was created which grouped spending patterns based on five different categories of purchases – two of which were considered fraudulent. The network was trained using only the genuine spending categories and then tested with data from all five categories. The classification was performed by comparing RMSE < 0.05 for legal transactions and RMSE > 0.18 for fraudulent transactions. It was found that the system had a 100% accuracy rate in detecting legal transactions, and an 85% accuracy rate for detecting fraudulent transactions.

Support vector machines were tested against a German credit card dataset from the UCI repository [9]. The dataset had 20 attributes and class labels 'G' and 'F' to represents genuine and fraudulent transactions, and there were 300 fraudulent transactions versus 700 genuine transactions available. It was found that the generalization capability of the SVM was mostly dependent on the type of kernel function that was used (linear, polynomial, or radial-based functions). The best results for each of these kernel functions were obtained using a v parameter set to 0.1, allowing for a more conservative boundary around the data points. It was concluded that the one-class SVM had a superior generalization ability compared with a binary SVM, with polynomial kernels achieving 92% accuracy.

A Bayesian network was used to detect fraud for data described by four features and a fraud label [16]. After learning, the topology of the network found that two features influenced fraud, while fraud influenced the remaining two features. To determine the fraudulent transactions, a cutoff threshold needs to be set to determine when transactions should be classified as fraudulent versus genuine. It was found that when 68% of fraudulent transactions are correctly classified, only 10% of genuine transactions are incorrectly classified. The same experiment was repeated for a dataset with 10 features, and it was found when 15% of fraudulent transactions were incorrectly classified, 73% of fraud transactions were identified.

FraudMiner was proposed as a frequent itemset approach for detecting credit card fraud [18]. In this approach, a legal transaction and fraud transaction pattern was determined by separating the legal and fraud transactions for each customer. The largest frequent itemset with the largest support was used as the rule for each classification. During testing, a matching algorithm was used to asses which profile the new transaction belonged to. It was found that the detection rate was 80% when considering 1400 customers, higher than any of the other algorithms involved in the experiment. The false alarm rate was also much lower than the other methods. Figure 9 shows the fraud catching rate using the FraudMiner as compared to some traditional outliers' detection methods.

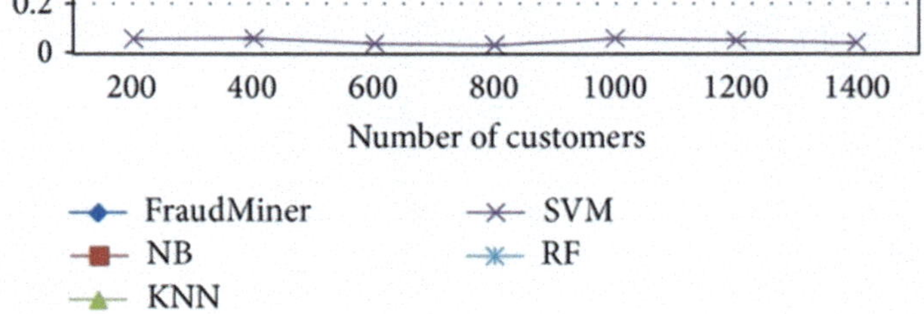

Fig. 9 Sensitivity/fraud catching rates by FraudMiner [18]

6 Conclusion

The traditional approach to outlier detection has been viewed as a one-dimensional problem. In this paper, we have categorized the outlier detection problem into two different types: (1) quantitative outlier detection, and (2) semantic outlier detection. We have shown that quantitative outlier detection is performed independently from the context of where the data is collected from. Alternatively, semantic outlier detection allows us to define the meaning behind an outlier based on its context, and train an algorithm to identify a specific class within the data. For credit card fraud, semantic outlier detection allows us to label fraud as the type of transaction to be identified as an outlier.

References

1. Agrawal R, Srikant R (1995) Mining sequential patterns. In: Proc. of the 11th Int. Conf. on data engineering, pp 3–14. https://doi.org/10.1016/j.jbi.2007.05.004.
2. Aleskerov E, Freisleben B, Rao B (1997) Cardwatch : a neural network based database stern for credit card fraud detection. Proc IEEE/IAFE:220–226. https://doi.org/10.1109/CIFER.1997.618940
3. Bhaduri K, Matthews BL, Giannella CR (2011) Algorithms for speeding up distance-based outlier detection. In: Proceedings of the 17th ACM SIGKDD international conference on knowledge discovery and data mining, pp 859–67. https://doi.org/10.1145/2020408.2020554
4. Breunig MM, Kriegel H-P, Ng RT, Sander J (2000) LOF: identifying density-based local outliers. In: Proceedings of the 2000 ACM Sigmod international conference on management of data, pp 1–12. https://doi.org/10.1145/335191.335388
5. Cooper GF, Herskovits E (1992) A Bayesian method for the induction of probabilistic networks from data. Mach Learn 9(4):309–347. https://doi.org/10.1023/A:1022649401552
6. Dau HA, Ciesielski V, Song A (2014) Anomaly detection using replicator neural networks trained on examples of one class. Simul Evol Learn:311–322. https://doi.org/10.1007/978-3-642-10439-8_15
7. Duan L, Xu L, Liu Y, Lee J (2009) Cluster-based outlier detection. Ann Oper Res 168(1):151–168. https://doi.org/10.1007/s10479-008-0371-9
8. Hawkins DM (1982) Identification of outliers. Fresenius' Z Anal Chem 311. https://doi.org/10.1007/BF00635536
9. Hejazi M, Singh YP (2013) One-class support vector machines approach to anomaly detection. Appl Artif Intell 27(5):351–366. https://doi.org/10.1080/08839514.2013.785791
10. Jiang F, Sui Y, Cao C (2011) A hybrid approach to outlier detection based on boundary region. Patt Recogn Lett 32(14):1860–1870. https://doi.org/10.1016/j.patrec.2011.07.002
11. Johnson T, Kwok I, Ng R (1998) Fast computation of 2-dimensional depth contours. Am Assoc Artif Intell 604:224–228
12. Knorr EM, Ng RT (1998) Algorithms for mining distance-based outliers in large datasets. In: 24th international conference on very large data bases, pp 392–403
13. Laskov P, Schäfer C, Kotenko I, Müller K-R (2004) Intrusion detection in unlabeled data with quarter-sphere support vector machines. PIK 27(4):228–236. https://doi.org/10.1515/PIKO.2004.228
14. Lei D, Zhu Q, Chen J, Lin H, Yang P (2012) Information engineering and applications, vol 154. https://doi.org/10.1007/978-1-4471-2386-6

15. Moore A, Wong W (2003) Optimal reinsertion: a new search operator for accelerated and more accurate Bayesian network structure learning. In: ICML, pp 552–559. http://www.aaai.org/Library/ICML/2003/icml03-073.php
16. Sam Maes, Tuyls K, Vanschoenwinkel B, Manderick B (1993) Credit card fraud detection using Bayesian and neural network. Interactive Image-Guided Neurosurguery 2:261–270
17. Schölkopf B (2002) Learning with kernels. J Electrochem Soc 129(November):2865. https://doi.org/10.1198/jasa.2003.s269
18. Seeja KR, Zareapoor M (2014) FraudMiner: a novel credit card fraud detection model based on frequent itemset mining. Sci World J 2014(August):252797. https://doi.org/10.1155/2014/252797
19. Shahid N, Naqvi IH, Qaisar SB (2015) One-class support vector machines: analysis of outlier detection for wireless sensor networks in harsh environments. Artif Intell Rev 43(4):515–563. https://doi.org/10.1007/s10462-013-9395-x

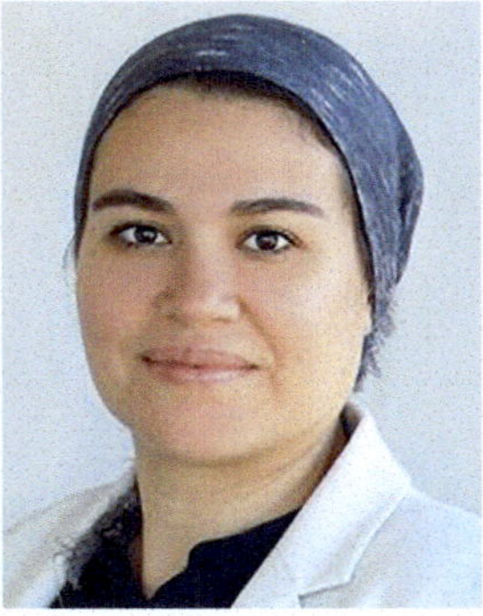

Dr. Rasha Kashef received her Ph.D. from the University of Waterloo, Department of Electrical and Computer Engineering in 2008. She is a professional engineer in Ontario. She worked as Assistant Professor at the School of Computing at the AAST Institute in 2009–2011. She also worked as a Research Associate at Microsoft Corp. Her research interests span the use of machine learning in big data analysis in different applications including healthcare, revenue management, and software engineering.

She worked as a postdoctoral fellow in the Department of Applied Mathematics at the University of Waterloo from 2011 until 2013. She also joined the Department of Management Science at the University of Waterloo from 2013 to 2016. She had been hired as an Assistant Professor at the Ivey Business School in Management Science group with a focus on Data Analytics from 2016 to 2019. She is currently an Assistant Professor in the Department of Electrical, Computer, and Biomedical Engineering at Ryerson University.

Michael Gencarelli is with the Ivey Business School, London, ON, Canada. He can be reached at mgencarelli.msc2018@ivey.ca.

Ahmed Ibrahim is with the Computer Science Department, Western University, London, ON, Canada. He can be reached at aibrah64@uwo.ca.

Cognitive Artificial Intelligence Countermeasure for Enhancing the Security of Big Data Hardware from Power Analysis Attack

Septafiansyah Dwi Putra, Arwin Datumaya Wahyudi Sumari, Adang Suwandi Ahmad, Sarwono Sutikno, and Yusuf Kurniawan

Abstract Digital communication systems as the part of big data are utilized to transmit data and information. The increase of the digital communication system utilization will increase the value of information and on the other hand also induces an increase in the number of attacks on such systems. Side Channel Attack (SCA) is an attack model that could disrupt the information security when hardware implements a cryptographic algorithm. Differential Power Analysis (DPA), a kind of SCA, can reveal 75% of secret key used in encryption hardware. Other techniques called Correlation Power Analysis (CPA) which uses correlation factor between trace and hamming weight from the input of key generation can reveal the right secret key of Advanced Encryption Standard (AES) in significantly shorter span of time. The objective of this research is to design and implement an electronic countermeasure to deal with power analysis attack. The attacking aspect is reviewed as a form of identification of the correct countermeasure method against power analysis attack using Cognitive Artificial Intelligence (CAI)'s method called cognitive countermeasure approach in an AES encryption device. Our main contribution is in the

S. D. Putra
Informatics Management, Politeknik Negeri Lampung,
Kota Bandar Lampung, Lampung, Indonesia
e-mail: septa@polinela.ac.id

A. D. W. Sumari (✉)
Department of Electrical Engineering, State Polytechnic of Malang,
Malang, East Java, Indonesia

Faculty of Defense Technology, Indonesia Peace and Security Center (IPSC),
Indonesia Defense University, Sentul, West Java, Indonesia
e-mail: arwin.sumari@polinema.ac.id

A. S. Ahmad · S. Sutikno · Y. Kurniawan
Cognitive Artificial Intelligence Research Group (CAIRG), School of Electrical Engineering and Informatics, Institut Teknologi Bandung, Bandung, West Java, Indonesia
e-mail: ssarwono@stei.itb.ac.id; yusufk@stei.itb.ac.id

Z. Md. Fadlullah, A.-S. Khan Pathan (eds.), *Combating Security Challenges in the Age of Big Data*, Advanced Sciences and Technologies for Security Applications, https://doi.org/10.1007/978-3-030-35642-2_4

design of cognitive-countermeasure by altering the measured power consumption in affecting the secret key value of power analysis. The measured signal is altered by generating random masking value using CAI's information fusion. CAI is a new perspective in Artificial Intelligence which is characterized by its capability to grow new knowledge based on the information from the sensory system. The random alteration of measured signal and continuous evolution of the masking value by using CAI's information fusion is very significant in tackling the risk of power analysis. We also succeeded in implementing an AES encryption device based on CAI method on the Field-Programmable Gate Array (FPGA) platform.

Keywords Big data · Cognitive Artificial Intelligence · Cognitive-countermeasure · Encryption · Field-Programmable Gate Array · Hardware attack · Information fusion · Information security · Knowledge-growing system · power analysis attack

1 Introduction

Digital communication systems as the part of big data system have become essential part of modern life. Information and data are transmitted electronically through digital-based communication systems. Various methods of maintaining information so that it is safe from tapping and theft in digital communication become more complex. The value of information will certainly increase when business needs an increase. On the other hand, the number of threats has increased as well. So that, the security aspect is to be critical in various information exchange transaction activities not only for the software but also for the hardware [1]. Security on hardware has become a very important metric that is needed to be considered among cost, performance, and power consumption used. Various forms of information prevention methods from attacks in the form of tapping and theft of information on digital communications encourage the use of several forms of cryptographic algorithms [2].

Cryptography is the science and technique of protecting information. A cryptographic algorithm is a function that uses a secret key to hide information. Without the knowledge of the secret key, the decryption processes of an information will be impossible. In general, the attacks on cryptographic algorithms concentrate on the mathematical aspects of the cryptographic algorithms. Designers assume that if the cryptographic algorithm is safe, then its implementation will also secure. This paradigm changed when Kocher published his research about timing attack [3], and attack based on power analysis [4]. Electronic devices that implement a cryptographic algorithm with mathematical assumptions can leak important information.

Side Channel Attack (SCA) is one model of attacks that can disrupt information security when a cryptographic algorithm is implemented on a hardware device [5, 6]. Differential Power Analysis (DPA) is one type of SCA that can reveal confidential information [7, 8]. Confidential information is secret information key used in cryptographic algorithms. The process of disclosing classified information is obtained by analysing various information leaks. The results show that by using DPA technique a 48-bit secret key value of 64 key bits can be revealed or the success rate is 75%. The attack technique produces 75% of all secret keys, while the remaining bits or 25%, are obtained with brute force techniques for Data Encryption Standard (DES) algorithm. The second technique has been proposed in various research namely using a correlation factor between the trace and hamming weight of the data being processed [9]. In some previous studies, the subkey of the secret key from Advanced Encryption Standard (AES) and DES cryptographic algorithms have been obtained with a large number of trace [10, 11]. When the number of traces used for the attack is too low the results get wrong. But when increasing the number of (big data of traces) the answer gets correct. With all the hardware security threats that happen here, intruder and cryptanalyst are fortunate to have big data analytics step in to help them. It will improve the reveal confidential information such SCA. There are several ways in which this is done by identifying changing use traces patterns and performing complex correlations across huge data traces.

The previous DPA attack pattern uses a lot of trace resources (>1000 traces) to get 75% of the correct bit value from the master key [12]. Improvements from the previous attack model were found when factors were trace and hamming the weight of the data processed is correlated. However, the assessment of attack correlation must have the ability to fully control the plaintext value which will be encrypted on cryptographic devices [13]. In addition to study the aspects of the attack, researchers have previously performed various forms of countermeasures software and hardware level for SCA-type attacks. Such methods include mechanical countermeasures on software level which are divided into transforming and masking data [8], and handling the form of hardware in the way of desynchronizing and noise generators [14–16]. Common weaknesses encountered in countermeasure aspects is between the performance and cost values that are considered not suitable in the embedded environment system.

At present, the latest approach for solving various problems related to cybersecurity with Artificial Intelligence (AI) has been found. The method describes how AI can perform the process of combining information, computing, and procedures in adapting the solution to a problem. In other research, Cognitive Artificial Intelligence (CAI) as a new perspective in AI has also been developed [17] and it has a big prospect to deliver solution regarding to problems related to cybersecurity. CAI is built based on the observation on how human brain grows new knowledge. The growing of knowledge is essentially based on information-inferencing fusion which it essentially combines information perceived by the sensory system when performing observations to the environment as the time passes. Information-inferencing fusion is the fundamental mechanism for gaining new knowledge. Information-inferencing fusion is a new and relevant technique for getting information (masking) value to provide decision alternatives which are the results from the fusion of a lot of information from multiple sources. This method is needed to cope with the fast dynamic changing occured in the environment.

In this research, a new masking technique based on CAI, called as cognitive-countermeasure is introduced. This technique uses random values generated through the incorporation of information-inferencing fusion. Our proposed technique is followed by its design and implementation on Field Programming Gate Array (FPGA) platform.

2 Research Objectives

The primary purpose of the research is the design and implementation of a reliable CAI-based countermeasure method for coping with the power analysis-based attacks on the AES encryption algorithm. Apart from these primary objectives, this research also aims to:

1. Produce a Device Under Test (DUT) test device for AES encryption on FPGA platform
2. Identify the vulnerabilities of AES encryption devices after a power analysis-based attack is carried out
3. Produce a cognitive-countermeasure method by generating precision and dynamic masking values. Changes in the form and the power consumption are carried out by using CAI's information fusion. Changes in the structure and the consumption of power in a precise and dynamic manner are considered significant in overcoming the risk of attack-based power analysis
4. Produce an AES encryption device on the FPGA platform based on cognitive-countermeasure method

3 Problem Solution

This section discusses the process in the design, implementation, and examine of the cognitive-countermeasure model on AES encryption device which is built on FPGA platform.

3.1 Use of CAI as DPA Countermeasures

One phenomenon that occurs in the human intelligence system is the ability of human brain to increase its knowledge by learning new things that occur around human. This phenomenon is then called as knowledge growing. Based on this phenomenon, a system that can emulate the mechanism of human brain in increasing human knowledge has been developed. This system is called as Knowledge Growing System (KGS) as depicted in Fig. 1, which was developed by Arwin Datumaya Wahyudi Sumari and Adang Suwandi Ahmad in 2009. In the brain, inputs regarding a phenom-

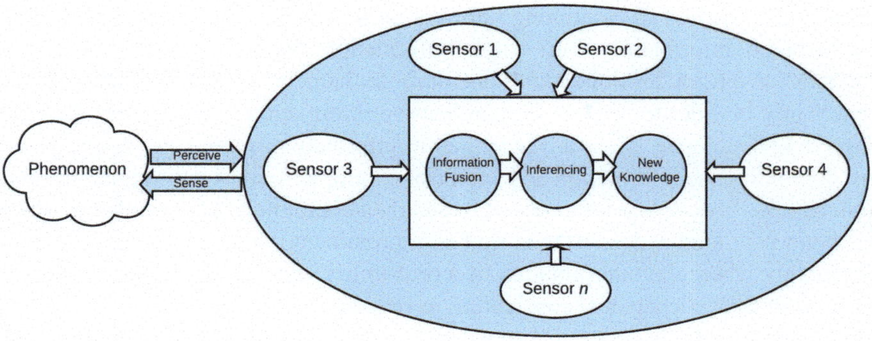

Fig. 1 The basic concept KGS

enon in the form of information from each human sensory then undergo a process to produce new information that is used as the knowledge for making decisions or actions. Furthermore, the perceived information is combined with prior knowledge, if already available, to produce more complete knowledge (posterior knowledge).

The obtained knowledge is then compared to the knowledge that has been previously stored in the brain to produce a conclusion (inferencing) about the observed phenomenon perceived by the sensory system. This conclusion will be the basis for making decisions or actions to be taken to the phenomenon. Based on the information-inferencing mechanism algorithm on the way of human thinks, a mathematical modelling is devised to emulate the mechanism of knowledge growing in KGS. This is done by assuming that the sources of information, namely the human sensory organs, observe the same phenomenon and each has its own information about it. The number of combinations of information delivered by the five sources of information can be calculated using Eq. 1 [18].

$$\lambda = \left(2^{\delta} - \delta\right) - 1 \tag{1}$$

λ is the maximum number of fused information or Guesses of the Occured Phenomenon (GOP), and ũFF; is the number of the sources of information in this case ũFF; = 5, represents the number of sensor in the human sensory system, namely eyes, nose, ears, tongue, and skin. Thus, the maximum number of OGP will be as much as λ = (25–5) − 1 = 26 based on the number of fused information will be obtained after the computation, which is the knowledge of KGS. KGS which is the foundation of CAI, is a system which emulates the way of human brain generateds or constructs or grows new knowledge. The knowledge will continue grow as the accreation of information received for the sensory system as the time passes. This mechanism represents the way of human learns from childhood to adulthood of something new by not throwing away the old already learnt, and enriches the already-stored knowledge.

Combining a lot of data from various sensors is naturally done by living things to assess the accuracy of the state of the surrounding environment and identify threats, thereby increasing their chances of survival [19]. Information fusion is defined as a method of combining information from many sources of information (multisource information) into single information. Various information fusion methods have been developed with one purpose, namely, to deal with information uncertainty in order to produce comprehensive information in order to explain the observed phenomena. More advance than it, information-inferencing fusion is not just combining the information delivered from multiple sources into single information, but also making the inference of it, and then combining the multiple inference to obtain new knowledge. BIM is said as a prominent paradigm for dealing with uncertainty where the interpretation of a probability value is the degree of trust of the subject or one's personal considerations based on experience [20].

In some way, humans acquire knowledge in probabilistic way. The thought process involves creating GOP automatically followed by finding and selecting the most appropriated answer from all potential possible GOPs. When thinking, the human brain makes GOPs which are considered as possible answers to what is being thought. In the context of AI, the inference method which mostly used is BIM in which the method is based on Bayesian rules as defined by Eq. (2).

$$P\left(B_j|A\right)=\frac{P\left(A|B_j\right)P\left(B_j\right)}{\sum_j P\left(A|B_j\right)P\left(B_j\right)} \tag{2}$$

In the above formula, notation $P(B_j| A)$ means the chance that GOP B_j is occured given information A, while $P(A| B_j)$ is prior chance that information A is true given GOP B_j, $P(B_j)$ is prior chance that GOP B_j is true, and $P(A) = \sum_j P(A| B_j)P(B_j)$. The inference extracted from the results of BIM computation can be obtained by applying Maximum A Posteriori (MAP) Technique or Point Estimation as shown by Eq. (3).

$$P\left(B_j|A\right)_{estimate}=\max_j\left(\frac{P\left(A|B_j\right)P\left(B_j\right)}{\sum_j P\left(A|B_j\right)P\left(B_j\right)}\right) \tag{3}$$

In reality, information received by human sensory system is not only A but also A_i, which has a consequence $P(B_j| A)$ becomes $P(B_j| A_i)$. BIM + MAP formula cannot cope with this situation, namely means the chance that GOP B_j is occured given information A_i, information from multiple sources. This one is called as multiple information multiple GOP situation. To apprehend the phenomenon such this, Arwin Datumaya Wahyudi Sumari and Adang Suwandi Ahmad perfected BIM + MAP formula by devising a new formula called as Maximum Score of the Total Sum of Joint Probabilities (MSJP) [21] which then renamed as ASSA2010 (Arwin Sumari-Suwandi Ahmad 2010) [22], as the core engine of KGS. ASSA2010 is presented in Eq. (4).

$$P\left(\psi_1^j\right) = \frac{\sum_{i=1}^{\delta} P\left(\upsilon_i^j\right)}{\delta} \tag{4}$$

where $P\left(\psi_1^j\right) \in \Psi$ is called as New Knowledge Probability Distribution (NKPD) at observation time, γ_1. The inferencing or the new knowledge at this point can be obtained by applying Eq. (5).

$$P\left(\psi_1^j\right)_{estimate} = \odot\left[P\left(\psi_1^j\right)\right] \tag{5}$$

where $\odot[\ldots] =$ max [...].

In principle, KGS is taking various veracity of the information to produce novel knowledge. In this research, KGS is used as a tool for creating randomized signal, i.e. take false information to produce a falsehood of information that will be used to make deception to be thought as a form of signal leakage by the attacker. Simply put, when applied to a cryptographic device, KGS can be used as SCA cognitive-countermeasure so that such devices become intelligence to its surrounding environment, be able to analyse and draw conclusions from information that has been collected by its sensory system, then take action based on the knowledge it obtaines. KGS implementation on cryptography device can be a solution to develop SCA cognitive-countermeasure. Equipped with KGS algorithm, cryptographic device can be used as the main control of an autonomous system and performs randomization signal trace, which it has ability to grow its own knowledge continuously, as the time passes. With such cognitive system, an early trace analysis for any possibility of disaster will be possible.

3.2 The Design of Device under Test Against DPA Attack

AES is the de facto standard worldwide since 2001 where National Institute of Standards and Technology (NIST) had chosen Rijndael block cipher as the new encryption standard [23]. AES has excellent performance on various forms of application and platform. Several approaches to software and hardware design for AES is continuing to be developed. The requirements in the context of throughput, strength and compact design has challenged the designers to fulfil those ones. Some of the best approaches in the AES optimization design are generally divided into two things, namely:

- Algorithm Optimization as DUT/CPA attacks: AES is based on finite field operations (finite fields). Furthermore, round AES has several interesting properties which makes possible to design the encryption and decryption process with the same technique. In our design, AES is built on the FPGA platform to produce efficient results from several AES rotation transforms

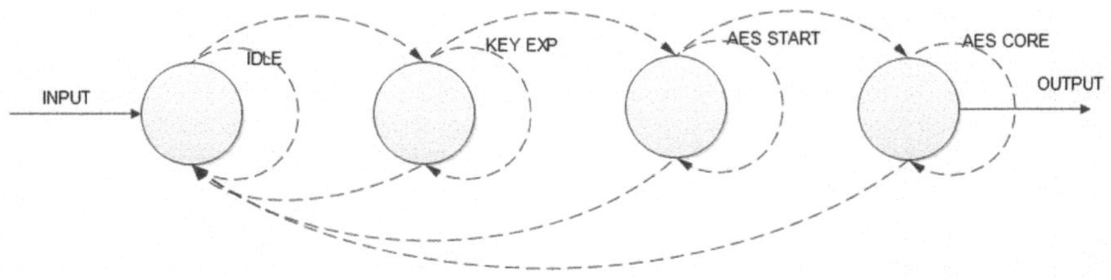

Fig. 2 Finite state of the AES machine [24]

- Architecture Optimization: In this DUT design, the standard architecture such as pipelining to improve hardware design throughput of AES is produced. For example, a systematic design approach for separating data and scheduling key generation is needed in producing the design efficiency of the AES DUT

The Finite-State Machine (FSM) for DUT device is shown in Fig. 2. The key generation process can produce the key needed to be processed with the secret spin key. The *clk* function is as input in key generation, the generated keys are stored in internal Read Only Memory (ROM) and are read by encryption and decryption blocks for each round. The encryption/ decryption module will accept 128-bit plaintext or ciphertext input when decryption is inactive (If $En = 1$ or 0 the encryption or decryption process).

The round operation performed on the DUT are 10 rounds by the form AES128 in general. In the first nine rounds of the encryption process, the cipher module will use *SubByte, ShiftRow, MixColumn,* and *AddRoundKey* operation. The final operation (round 10) is performed without *Mixcolumn* operation in completing one block of the encryption process. Figure 3 shows the structure of the AES128 algorithm that was implemented in the DUT.

The cipher module performs data encryption or decryption. In an AES algorithm with a 128-bit key, the cipher module does ten rounds of substitutions and permutations to encrypt the data input (plaintext). In the first nine rounds of the encryption process, the cipher module uses *SubByte, ShiftRow, MixColumn,* and *AddRoundKey* operations. In the final (tenth) round, *Mixcolumn* is used to complete the block encryption process. The process begins with the key generation module (*KEY_EXP*) which is the initial stage to provide input values to the round key module (*key_rnd*). After the key generation process is done, the *AES_CORE* module will work by configuring one round for one clock. The reason of giving one round to one clock is to speed up the encryption process and minimize the amount of clock usage which have implications for the cost value and the performance of the designed device. Therefore, the clock value per encryption process is 11 clocks and the frequency used in the DUT is worth $\left(\frac{1}{3.148}\right)ns$.

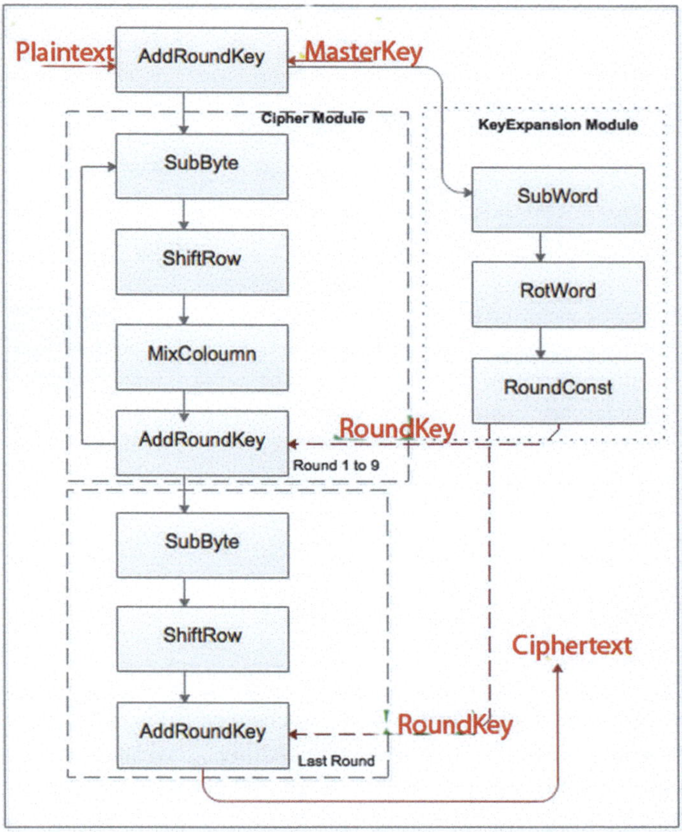

Fig. 3 DUT AES algorithm [24]

$$Throughput = \frac{128\,Bits\,x\left(\frac{1}{3.148}\right)ns}{11\,clocks} \tag{6}$$

Throughput obtained for each unit of time is 3.696 Gbps.

3.3 Power Analysis Attack against AES Encryption Device

In this section, the DPA/CPA attack on the AES encryption device that has been designed is carried out. The measurement technique with side channel declares a set of digital information that is converted from an analog physical leak at a sampling frequency. This contains a physical hole that has been digitalized against discrete time determined by several sample points. Side-channel measurement data also

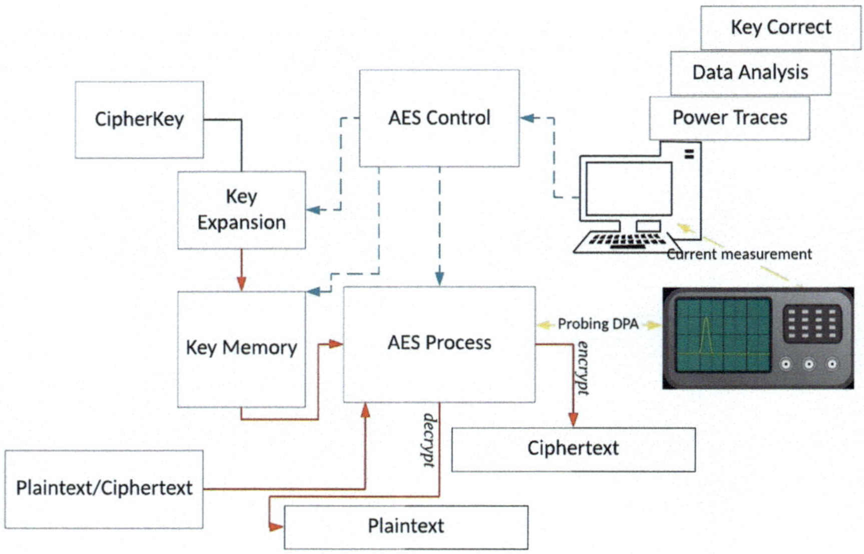

Fig. 4 Attacking AES encryption device with DPA/CPA [25]

called trace, like power trace in measuring electrical power consumption on a device. Trace information that is digitized and changed from analog physical leakage is significantly affected by measurement settings, environmental parameters, and other information factors. Therefore, trace must have noise measurement. Noise in power trace relatively small when compared to side-channel information. Figure 4 is an illustration of the architecture in measuring data from plaintext or ciphertext and power trace.

In this simulation, the attacks on the AES device is done by guessing the key used in the initial *AddRoundKey* operation, that is before entering the encryption round. The supposed key is 8-bit in size according to the contents of one state. Therefore, the iteration is carried out in order to try every possible key that might be used in one state. The test results shown in Table 1. Base from this simulation, the use of Difference-of-Means (DoM) techniques is considered less effective and less efficient compared to CPA techniques. Seen from the number of traces needed, the DoM attack technique requires more traces. The DoM computing time is longer than CPA techniques. In general it can be concluded that the weak point of an encryption device is when an attacker is able to estimate the value of Hamming Weight (HW).

In addition to the examination aspects of the DPA/CPA attack, this research produced several possible attack surfaces for attacking purpose. Attack surface of the DPA/CPA attack technique is the estimated value of power consumption (P_{hyp}) which has connection with data and power consumption. The following is the list P_{hyp} for the AES encryption algorithm in Electronic Code Book (ECB) mode.

The HW value is obtained by performing operations as follows:

Table 1 Simulation results of attack to AES encryption device

Testing variable	DPA/DoM results	CPA/Pearson results Correlation
Trace amount used	1050	500
Time spent on execution	960 seconds	123.2 seconds
Number of key bits obtained	128-bit	128-bit
Number of key bits that cannot be obtained	0	0

Table 2 Attack surface AES device

No.	Attack Surface	Information
1	$P_{hyp} = HW(SBox(P_i \oplus K_j))$	CPA *attack*
2	$P_{hyp} = HW(Sbox(P_i \oplus K_j)) \bmod 2$	DPA *attack*
3	$P_{hyp} = HW(Round9 \oplus Chipertext)$	DPA/CPA *attack from last round*
4	$P_{hyp} = HW(InvSbox(InvShifRow\ (R_{10} \oplus K_{10}) \oplus) R_{10})$	CPA *attack*
5	$P_{hyp} = LSB(SBox(P_i \oplus K_j))$	DPA *attack*

$$x = \left(x_1, x_2, \ldots, x_8\right) \text{where } x_i \in \{0,1\} \tag{7}$$

$$HW\left(x\right) = \sum_{i=1}^{n} x_i \tag{8}$$

Based on Table 2, the DPA/CPA attack is an attack which is obtained through the attacker's ability to search for the hypothesis of the power consumption (P_{hyp}) accurately. So that the most feasible approach is to decompose the power consumption as efficiently as possible. The deception of the power consumption will release the connection between data and power consumption of the encryption key used [26].

4 Countermeasure Design on Power Analysis Attacks

Based on the results of the previous tests, it has been obtained several attacks surface in carrying out a DPA/CPA attack from determining the estimated value of power consumption. The determination of the estimated cost of power consumption is the key success of a DPA attack. Therefore, to protect the estimation of the power consumption is to change the amount of power consumption. Dynamic changes in the most dominant factors will give a very dynamic phishing effect. One method that is used as a contribution and also the novelty in this research is the use of information integration approach (information fusion) in DPA/CPA countermeasure. Another novelty generated from this research is the improvement of the masking method produced by previous researchers (state of the art). This novelty is the search mechanism for precision masking values and in a limited amount of masking. The limitation on the number of masking aims to provide cost as little overhead as possible in the design aspects of the device.

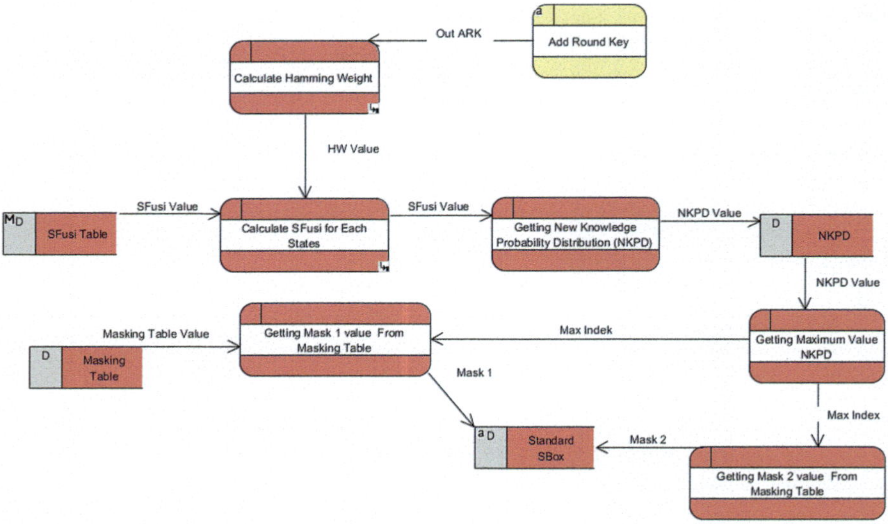

Fig. 5 Data Flow Diagram of AES architecture with cognitive countermeasure approach

The generation of masking numbers is based on the need of determining the right masking number and they must be generated quickly. A correct masking value is a masking value that has a significant impact and affects key correlations in DPA attacks. The association is the value between the key guesses by the attacker and the n-trace value generated by the encryption device.

Figure 5 shows the detail description of the mechanism for generating masking values and it is the main novelty of this research, namely CAI countermeasure. Specifically, the masking value generation system has subsystems in the form of sub-processes consist of *calculate Hamming Weight*, *calculate SFusi*, *getting NKPD*, *getting the maximum value of NKPD*, *getting Fmask 1*, and *getting Fmask 2*. The sub-processes are used for data storage in the form of registers for *SFusi, masking,* NKPD, and *SBOx standards*. The *calculate Hamming Weight* sub-process receives input from the output of *AddRoundKey* sub-process. The output of the sub-process *calculate Hamming Weight* is the amount of HW value that will become the input for the next sub-process. The *calculate SFusi* sub-process will produce NKPD from the input values of the *SFusi* table register.

The proposed countermeasure architecture given in Fig. 6 is based on CAI's information fusion. Each input value given in information fusion part is a value that influences the decisions that will be produced in the future. The first step of the cognitive-countermeasure system is to receive fusion data from the Sensors $(S_1 ... S_{16})$ which in this research are represented as output from *AddRoundKey* and *Hypothesis* $(H_1 ... H_8)$ in this research is mentioned as hamming value weight ([0..8]) that has an effect. The best hypothesis of each sensor value is the value obtained in the *SFusi* table. The correspondence value from *SFusi* will be entered and processed to become the

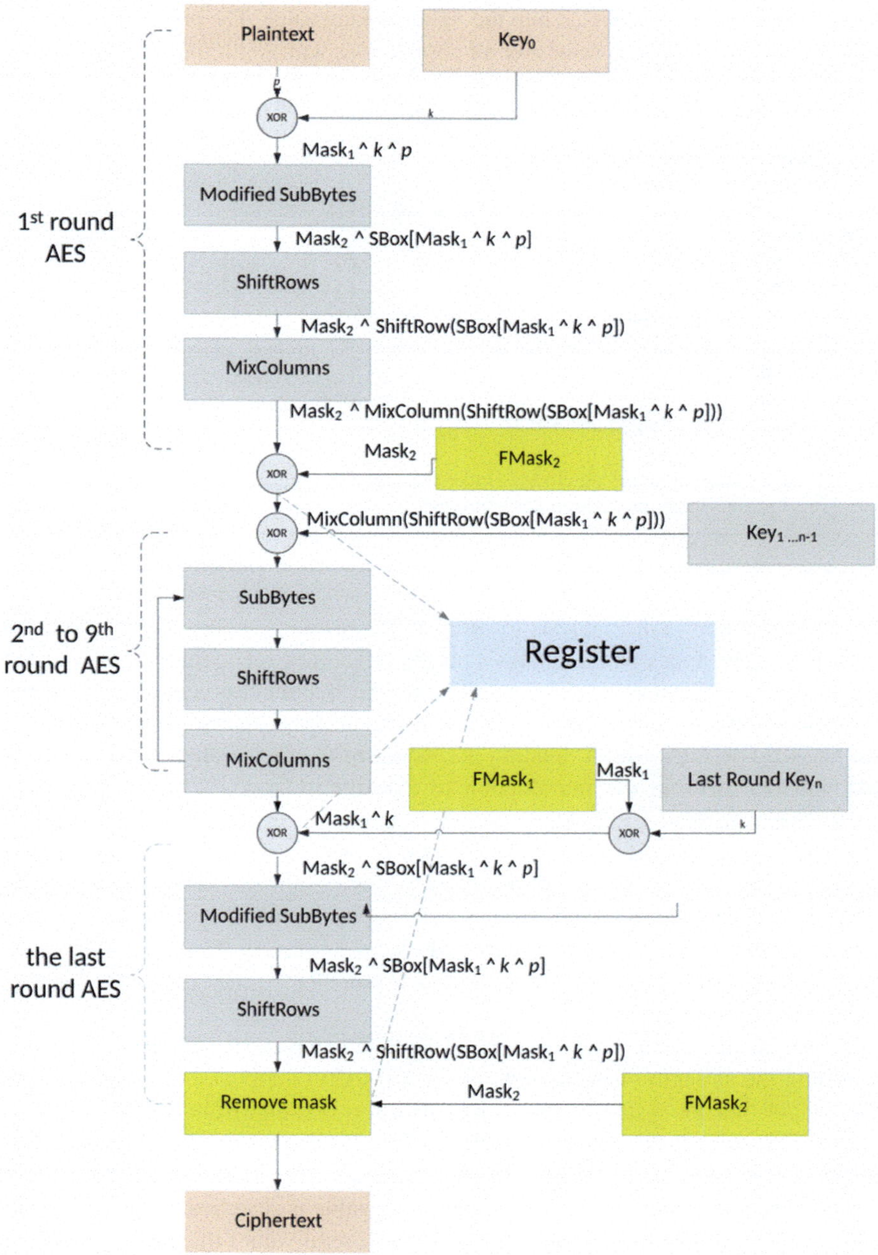

Fig. 6 AES architecture with cognitive countermeasure approach

NKPD values to be reprocessed into the right masking value. The process of calculating the best hypothesis (*H*) values for all observations with Algorithm 1 as follows.

Algorithm 1. Cognitive Masked Function	**Cost**	**Times**	**Big O Notation**
Input: Output_AddRoundKey and	C_1	*1*	
Output: Fmask1, Fmask2			
0: State = HammingWeight (Output_Ad-	C_2	*1*	*O(1)*
dRoundKey)	C_3	*4*	*O(4)*
1: **for** i= 0 to 4 do	C_4	*4*	*O(4)*
2: **for** j= 0 to 4 do	C_5	*1*	*O(1)*
3: temp = *state*(i,j);	C_6	*1*	*O(1)*
4: NKPD[i,j] = SFusi (temp);			*O(1)*
5: **end**		*1*	
6: **end**	C_7		*O(16)*
7: Indexs= maximum(sum(NKPD));		*1*	*O(1)*
8:	C_8	*1*	*O(1)*
10: Fmask1 = randomValueIndexs(Indexs)	C_8		
11: Fmask2 = randomValueIndexs(Indexs)			
			Total = O(42)

Big O notation is the analysis to estimate the resource requirements to run the algorithms. The focus of the analysis are the resources for memory cost and the amount of the computational implementation. The amount of processing time is the execution time for each instruction. In the second analysis, the worst case is used as the basis for writing Big O notation according to Cormen's algorithmic analysis technique in 2011 in the introduction to algorithms book [27]. The cognitive-countermeasure design is divided into two components. In algorithm 1 the value of time and cost is $T(n) = C_1 + C_2 + C_3 4 + C_4 4 + C_5 + C_6 + C_7 + C_8 2$ with the amount of notation O (42) = O (1). In algorithm 2, this design has several notations O (255) = O (1) in the modified *SBox* generation operation. But theoretically, this design is fixed both in time complexity and space complexity. That causes the number of overhead areas to have a much smaller value than the others' countermeasure-algorithm.

Besides being based on the analysis of the time complexity, other considerations regarding the strength of the aspects of the DPA/CPA attack are discussed. In the above algorithm, the system receives input in the form of value from the *AddroundKey* output. This value is processed to obtain the *Fmask1* value from *AddRoundkey* outcomes, and *Fmask2* is masking for *SubBytes* function. The input value in the system is represented in binary units, and the hamming value is searched weight it is. HW value will be obtained when the corresponding several values that are not worth 0. After getting the HW value, the next step is to represent the HW value from each state into the information fusion function.

Information fusion tables, as shown in Table 3 as an example, are the tables used in transforming HW values into processing tables. After getting the values from the

Table 3 *SFusi Table*

	H_1	H_2	H_3	H_4	H_5	H_6	H_7	H_8
HW_0	0	0	0	0	0	0	0	0
HW_1	1	0	0	0	0	0	0	0
HW_2	0	1	0	0	0	0	0	0
HW_3	0	0	1	0	0	0	0	0
HW_4	0	0	0	1	0	0	0	0
HW_5	0	0	0	0	1	0	0	0
HW_6	0	0	0	0	0	1	0	0
HW_7	0	0	0	0	0	0	1	0
HW_8	0	0	0	0	0	0	0	1

information fusion part for a 16-byte matrix, the HW value that most influences the observation will be searched. To protect the *SubBytes* transformation thoroughly, all values in the *SBox* are recalculated. The process of repeating the *SBox* computing process for each encryption process is carried out by the mechanism shown in Algorithm 2.

Algorithm 2. Cognitive Countermeasure SubBytes	**Cost**	**Times**	**Big O Notation**
Input: N *Output_AddRoundKey* and F_{mask1}, F_{mask2} = *generated from information fusion*	S_1	*1*	S_1
Output: *maskingSBox*			
1: **for** *i*= 0 to 255 **do**			
2: *MaskedSBox*[i $\oplus$ F_{mask1}] = *SBox* [i]	S_2	*255*	*O(255)*
$\oplus$ F_{mask2});	S_3	S_3	*O(1)*
3: **end**			
			Total = O(256)

To perform the transformation *SubBytes* with cognitive-countermeasure *SBox*, each output value will be added *AddRoundKey* XOR function with the value of his *Fmask1* to each byte of the state matrix this time. After this nonlinear transformation, the final part will release the masking value *Fmask1* of the value sought. Through these two possible approaches, masking values *Fmask1* and *Fmask2* are added to the key scheduling process and will cover each round of AES. Every operation of this round key is done, it will include the actual state value in such a way that we need it. But we must calculate the expanded key every time, we change the mask for the *SBox*. In the next section, the process in more detail is explained, namely in other linear transformations. Other linear transformations are *ShiftRows* and *MixColumns* operations.

The form of shift-row transformation occurs in the second, three, and fourth lines as shown in Fig. 7. Value changes only happen in those lines. The values in $S_{0,0}$ up to $S_{3,5}$ are obtained by looking at the output of the *SBox* previously produced. For example on $S_{1,1}$ = $SBox(k_{1,1} \oplus p_{1,1})$ and $S_{1,1}$ value to be $S'_{1,2} = SBox\left(k_{1,2} \oplus p_{1,2}\right) \oplus F_{mask2}$. When a value change is made in the *ShiftRow* function that is to be, hamming leak distance for the function to be as follows.

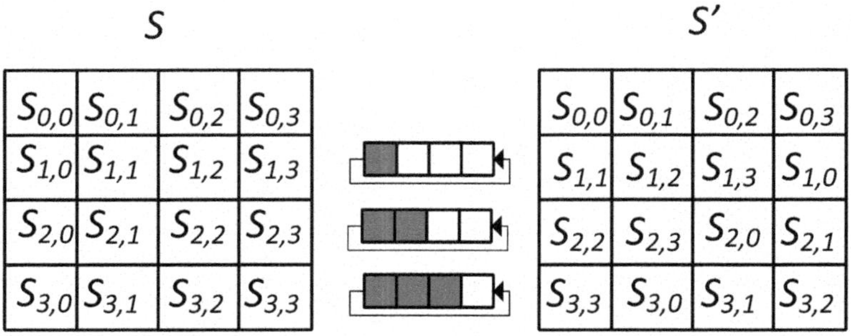

Fig. 7 *ShiftRow* function

$$HD\left(S_{1,1},S'_{1,2}\right)=\left(HW\left(SBox\left(k_{1,1}\oplus p_{1,1}\right)\right)\oplus HW\left(SBox\left(k_{1,2}\oplus p_{1,2}\right)\right)\oplus F_{mask2}\right) \quad (9)$$

In cognitive-countermeasure, changes in masking values give an impact on the changes in each state in the second, three, and four rows. The attacker can model the power consumption by making hypothesis (*H*) of it from each change of power by estimating at least $2^{16} = 65{,}536$ possible value combinations so that the value of the masking value in the *ShiftRow* function will be very difficult to be analysed for power attacks purpose. This change will also result in a more exponential value when the masking value is changed in each round and the encryption process. At least, in AES with a key length of 128 bits, the *H* value for estimating the power consumption requires $10 \times 2^{16} = 655{,}360$ times.

The *MixColumn* function will multiply each column of the input block state with the matrix where the multiplication operations are the same as the matrix multiplication in general. In the cognitive-countermeasure algorithm, the value of each *MixColumn* state is obtained from the previous value, namely the output of the *ShiftRow* function as follows.

$$S_{1,1}=SBox\left(k_{1,2}\oplus p_{1,2}\right)\oplus F_{mask2} \quad (10)$$

For all states in the *MixColumn* function, it applies as in the *MixColumn* matrix in NIST FIPS197 standard. However, in the final part, XOR was performed with two randomization values obtained from cognitive-countermeasure calculations. The value of c shows the value of each column of c = (0, 1, 2, 3). For each column, apply *MixColumn* as follows.

$$S'_{3,c}=\left(\{03\}\cdot S_{0,c}\right)\oplus S_{1,c}\oplus\left(S_{2,c}\right)\oplus\left(\{02\}\cdot S_{3,c}\right)\oplus Fmask2 \quad (11)$$

When the value changes are made to *MixColumn* function that is $S_{1,1}$ to be $S'_{1,2}$, hamming leak distance for the function is as follows.

$$HD\left(S_{1,1},S'_{1,2}\right)=HW\left(S_{0,1}\oplus\left(\{02\}\oplus S_{1,1}\right)\oplus\left(\{03\}\cdot S_{2,1}\right)\oplus S_{3,c}\right)$$
$$\oplus HW\left(S_{0,1}\oplus\left(\{02\}\oplus S_{1,1}\right)\oplus\left(\{03\}\cdot S_{2,1}\right)\oplus S_{3,1}\oplus Fmask2\right) \quad (12)$$

4.1 Examining the Performance of Cognitive-Countermeasure Method

In AES design with cognitive-countermeasure, changes in masking values give an impact on the overall change in the value of the state. The attacker can model the power consumption by making *H* consumption from each power change by making estimates of at least 4 bytes of the values in the masking table of cognitive-countermeasure is 40 pairs. The power *H* value is an accumulation of the masking value between the *ShiftRow* and *MixColumn* functions. So that applying 2^{32}–40 results in 4,294,967,256 possible combination values. Giving a masking value greatly gives a very significant change in *H* power in the *MixColumn* function. The test results on the four validations are shown in Table 4.

The AESVS test results through the Known Answer Test (KAT) test and Monte Carlo Test (MCT) have shown results that meet the NIST FIPS197 standard. In this section, the results of the AES design are cognitive-countermeasure based with state of the current art will be analysed and compared. This comparison begins with examining the level of complexity of processing and cost used in DPA/CPA countermeasure.

Table 4 Results of testing cognitive AES devices masking

Testing Compo-nent	The number of Iteration	Encrypt/ Decrypt Pass	Time Before Counter-measure	Time After Counter-measure	The improve-ment percentage
AESVS *VarKey test data for* ECB AES128 (KAT)	128	Yes	92,505 ps	98,900 ps	6.91%
AESVS KeySBox test data for ECB	21	Yes	15,465 ps	17,675 ps	14.29%
AESVS VarTxt *test data for* ECB	128	Yes	92,505 ps	102,505 ps	10.81%
AESVS MCT *test data for* ECB	128	Yes	28,044,345 ps	38,055,645 ps	35.69%

4.2 FPGA Implementation of Cognitive-Countermeasure Method

AES design with cognitive-countermeasure has been realized with a repetitive architecture using 8-bit data lines on the FPGA platform. In this section, the AES design without cognitive-countermeasure (DUT) and AES accompanied with cognitive-countermeasure is compared. The most dominant difference in the number of slice is caused by the addition of logic gate for information fusion functions on Lookup Table (LUT) of 450 slices which are obtained from the amount of data in the masking table of 48 x 8360 plus other utilities. Along with the addition of slice to the LUT, it also implies that the use of LUT logic is worth 450 slices. The increasing use of slice causes an increase in the component width by 25% compared to DUT (Table 5).

This design uses 8-bit-wide data path for each AES round operation, and the main key expansion operation is carried out sequentially. In a cognitive-countermeasure-based AES design, the operations are carried out in parallel for pieces of input data (plaintext/ciphertext) and different keys. The use of this technique significantly reduces the number of total cycles and increases throughput. This design successfully maintains the hardware area and keeps the power consumption low compared to the DUT model. This comparison is shown in Fig. 8. The AES floor planning design supports 128-bit keys and counts one round at a time in 16 clock cycles. This design consists of six components, namely cognitive-countermeasure unit (red), *AddRoundKey*, *SubByte*, *MixColumns*, *ShiftRow*, and key expansion units.

After getting the design, we synthesised it in FPGA design software. The use of power consumption of the cognitive-countermeasure-based AES design is obtained. The main focus on finding the value of power consumption in the model is to look at the percentage of overhead over costs of the countermeasure. Table 6 shows the comparison of the power use of power consumption when countermeasure was given to the AES design. The use of the Application Specific Integrated Circuit (ASIC) platform provides a small power consumption, because ASIC is specifically used for specific functions. Unlike FPGA, architectural design is used for general processing purpose.

Table 5 Comparison of proposed design methods with the study of state of the art

	Circuit Area (slice)			
Countermeasure	Cipher	Countermeasure	Total	Type Countermeasure
Cognitive-countermeasure	2643	882 (25.02%)	3525	Cognitive-countermeasure
With UDRPG	2335	2370 (49.8%)	4750	Hardware
PRNG with DRO	2100	35,000 (>100%)	37,100	Software
With DCRO	2335	1455 (38.39%)	3790	Hardware
With dummy instruction	1424	1491(51.14%)	2915	Logic and gate

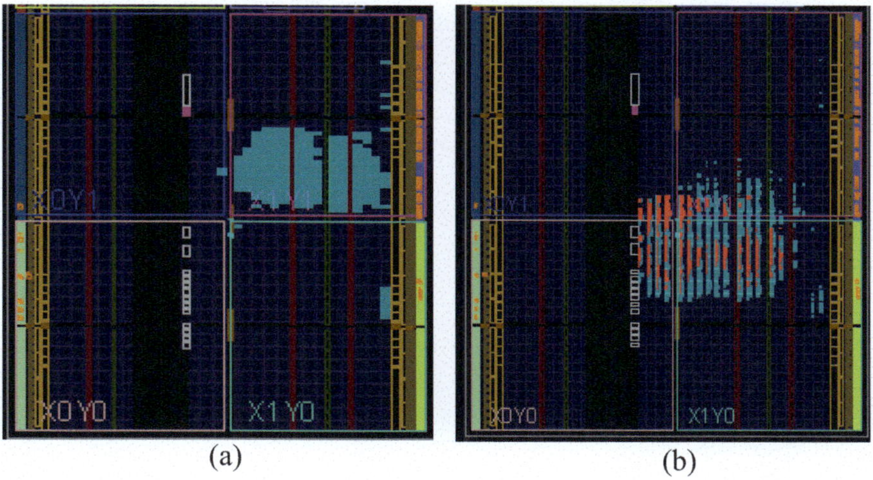

Fig. 8 Floorplanning in AES with and without cognitive-countermeasure approach (**a**) Without countermeasure (**b**) With countermeasure

Table 6 Comparison of proposed method power consumption with state studies of the art

Countermeasure	Circuit Power			Platform
	Cipher	Counter- measure	Total/ Overhead (%)	
Cognitive-countermeasure method	16.62 mW	2.38 mW	19 mW (14.3%)	FPGA
With UDRPG	68 μW	34.5 μW	102.5 μW (50.7%)	ASIC
PRNG with DRO	5.9 mW	1.11 mW	7,01 mW (18.8%)	ASIC
With DCRO	68 μW	44.5 μW	112,5 μW (65.4%)	ASIC
With dummy instruction	Not defined	Not defined	5%	FPGA

As shown in Table 6, the cognitive-countermeasure process requires high power consumption compared to other AES designs. The increase in power consumption is the consequence of adding the countermeasure mechanism. The consumption of 19 mW or an increase of 14.3% power use is the total accumulated consumption for carrying out key generation process, AES, Universal Asynchronous Receiver-Transmitter (UART) operations, and data storage in the registers. The simulation of the AES design attack based on cognitive-countermeasure also carried out key validation. The predicted keys are the 8-bit size for each state, so 8-bit will be searched for 16 states. The attack uses Algorithm 1, the same algorithm in examining the vulnerabilities in AES-DUT. The test began with the collection of 5000 pairs of traces. *SubKey* key search begins with finding the hypothesis of power value (*H*), which is doing a *bitxor* operation between the *H* key value and the plaintext used.

The output from *bitxor* operations it is the basis for finding the intermediate values. The next process is to calculate the mean value by using the value of the Least Significant Bit (LSB) as the intermediate value. After that, the simulation calculated the average value of the voltage model against the trace. The absolute maximum

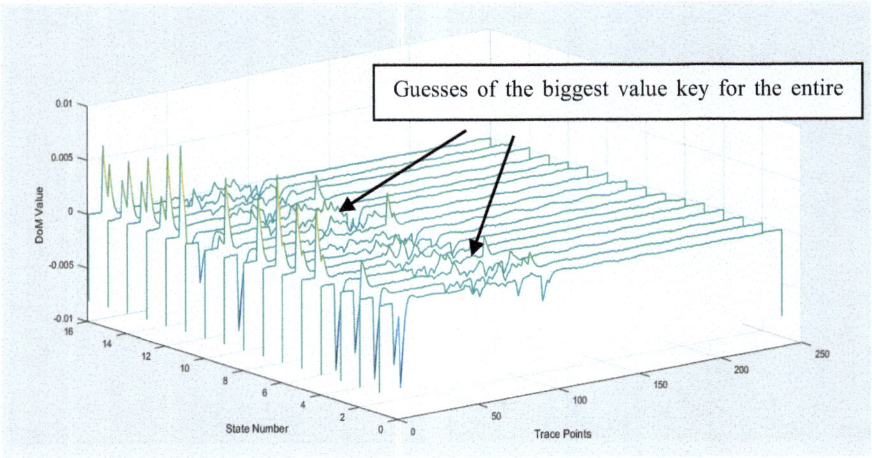

Fig. 9 Search results for AES sub-keys with DPA-DoM

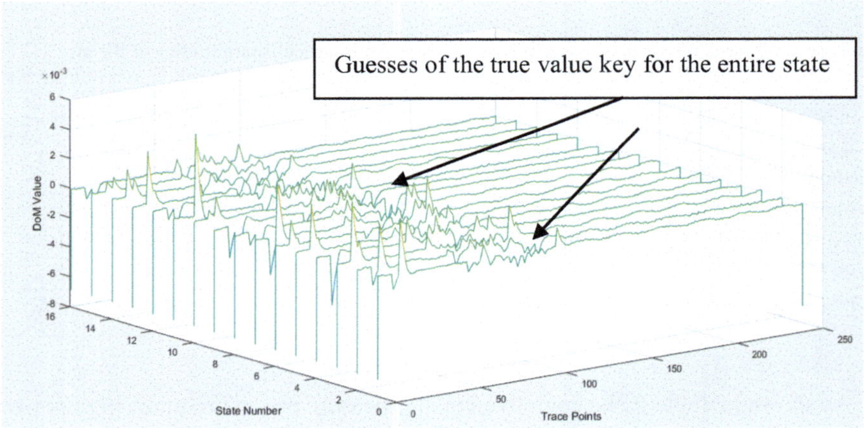

Fig. 10 The results of the curve traces for sub-keys are true

value of the average voltage against trace identifies the most appropriate key guess. Testing of DPA-DoM attacks on AES devices based on cognitive-countermeasure does not get a single sub-key that is the true value. The test results on keys that are considered correct are shown in Fig. 9.

Figure 9 shows the results of applying the most appropriate key to the AES sub-key as many as 16 states. Absolute maximum value of mean voltage against *trace* identified the most appropriate key guess is `9CDF1CFD9E98C54A7FB2C06D26AA6EE0`. The key that is the true value should be `2B7E151628AED2A6ABF7158809CF4f3C`. The *trace* curve for keys that are correctly valued is shown in Fig. 10.

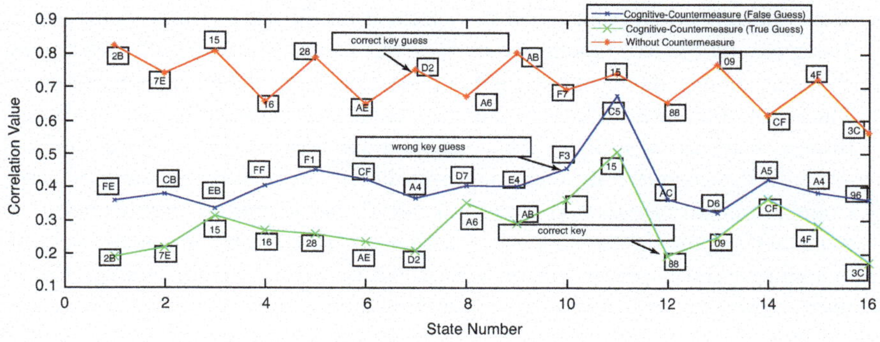

Fig. 11 Comparison of correlation values with and without cognitive-countermeasure

In Fig. 11, the absolute value for the entire state in the correct key is always below the absolute key value of the DPA-DoM attack. These results indicate that the use of AES encryption devices based on cognitive-countermeasure can protect against DPA-DoM attacks even though they use a few trace curves. On the software side, the researchers use MATLAB software to analyse and carry out CPA attacks. The CPA attack model focuses on attacks after the *SBox* value for the entire matrix AES state. Several bytes in the matrix AES128bit state amounts to 16 states. The attack model used in this test is the CPA method for the First AES HD Round proposed in [9].

5 Concluding Remarks

The conclusions and follow-up of this research that has been carried out are explained in this section. The further works are the next research plan that will be carried out by the researchers.

5.1 Conclusion

A DPA/CPA attack is a statistical attack based on an analysis of the power usage needed by the encryption device. Specifically, the conclusions of this research are as follows.

- The DUT AES128 on the FPGA platform as a testing environment for attack and anti-attack has successfully been designed. After analysing the DPA and CPA attacks on the DUT, the surface (point of vulnerability) of the AES encryption device has also been obtained.

- The main vulnerability of AES128 is located on the estimated value of power that is predictable as a function of its *SubBytes* in each round. Based on the point of vulnerability, the best model is obtained as a form of DPA-countermeasure. The countermeasure model is to calculate and make significant changes in power form on AES attack surfaces. This change in the form of power consumption is done by the cognitive masking approach based on cognitive-countermeasure. Cognitive-countermeasure is built on the analysis of the most significant power usage at one time the message encryption processed. Cognitive-countermeasure is a method developed from CAI's information fusion. The information fusion function is to combine information quickly and precisely to obtain the best masking value decision. The results of the tests show changes in the form and the power consumption randomly and precisely manner in overcoming the risk of power analysis attacks
- This research presents a new approach to countermeasure power analysis attacks efficiently to be implemented in hardware and can prevent such attacks. This research also presents attack statistics to calculate the probability that certain DPA/CPA attacks cannot be carried out. An AES encryption device has been produced on the FPGA platform with dynamic cognitive-countermeasure method. Cognitive countermeasure that is dynamically interpreted as randomizing the value of an AES secret key by giving a masking value which follows the message value. With this mechanism, the attacker is not able to search the key. The encryption device has been validated by various tests required by the NIST FIPS197 standard
- The results of the NIST FIPS197 test on the AES encryption device based on cognitive-countermeasure showed an increase in processing time by 16.97%. The validation results on CPA attacks show that cognitive-countermeasure decreases the correlation value after applying cognitive-countermeasure is 0.0508, previously the correlation value without masking is 0.8 with the number of traces 5000. Regarding the hardware resource usage, there is an increase in the sequence number (slice) amounted to 25.02% of the design of the early DUT as the consequence of adding the cognitive-countermeasure parts. But the percentage value is still smaller compared to the other four comparators studied in the state of the art.
- The AES encryption device based on the cognitive-countermeasure method has been built and is able to perform deception to the attacker in performing the key search. The deception is made possible by using the knowledge obtained from CAI approach. This design has higher reliability against various power analysis-based attacks (DPA/CPA).

5.2 Further Works

- With the increasing of hardware security threats nowadays, cryptanalysts are fortunate to have big data analytics step to reveal confidential information such as SCA (DPA/CPA) by performing complex correlations across huge data traces. To anticipate attacks such these, we found that the combination of CAI and cryptanalysis gave surprising results. Indeed, we found an improvement of the best DPA/CPA countermeasure namely cognitive-countermeasure. Our model seems to be more effective than the most effective countermeasure method against encryption device. In the design aspect, especially by testing NIST FIPS197 on the device shows an increase in processing time by 16.97% whereas the elements of the use of hardware resources increases to 25.02% of the designed DUT. The potential for future research that can be developed is how to reduce the processing time and cost overhead of the AES encryption device design based on cognitive-countermeasure
- The future research will also focus on the optimization of the use and the design of cognitive-countermeasure as a masking technique. This includes conducting the test and the design of DPA/CPA countermeasure on other cryptographic algorithms such as RSA, SHA1, and MD5

References

1. Hess E, Janssen N, Meyer B, Schutze T Information Leakage Attacks Against Smart Card Implementations of Cryptographic Algorithms and Countermeasures, p 10
2. Joux A (2009) Algorithmic cryptanalysis. CRC Press, Boca Raton
3. Kocher PC (1996) Timing attacks on implementations of Diffie-Hellman, RSA, DSS, and other systems. In: *Advances in Cryptology — CRYPTO '96*. Springer, Berlin/Heidelberg, pp 104–113
4. Kocher P, Jaffe J, Jun B, Rohatgi P (2011) Introduction to differential power analysis. J Cryptogr Eng 1:5–27
5. Messerges TS, Dabbish EA, Sloan RH (1999) Investigations of power analysis attacks on smartcards. In: Proceedings of the USENIX Workshop on Smartcard Technology on USENIX Workshop on Smartcard Technology, Berkeley, CA, USA, pp 17–17
6. Katashita T, Hori Y, Sakane H, Satoh A (2011) Side-Channel Attack Standard Evaluation Board SASEBO-W for Smartcard Testing, p 8
7. Akkar M-L, Bevan R, Dischamp P, Moyart D (2000) Power analysis, what is now possible. In: Okamoto T (ed) Advances in Cryptology—ASIACRYPT 2000, vol. 1976. Springer, Berlin/Heidelberg, pp 489–502
8. Golić JD, Tymen C (2003) Multiplicative masking and power analysis of AES. In: Kaliski BS, Koç ç K, Paar C (eds) Cryptographic Hardware and Embedded Systems – CHES 2002, vol. 2523. Springer, Berlin/Heidelberg, pp 198–212
9. Brier E, Clavier C, Olivier F (2004) Correlation power analysis with a leakage model. In: Joye M, Quisquater J-J (eds) Cryptographic Hardware and Embedded Systems – CHES 2004, vol. 3156. Springer, Berlin/Heidelberg, pp 16–29

10. May D, Muller HL, Smart NP (2001) Random register renaming to foil DPA. In: Koç ÇK, Naccache D, Paar C (eds) Cryptographic Hardware and Embedded Systems—CHES 2001, vol. 2162. Springer, Berlin/Heidelberg, pp 28–38
11. Shan W et al (2014) A side-channel analysis resistant reconfigurable cryptographic coprocessor supporting multiple block cipher algorithms. In: 2014 51st ACM/EDAC/IEEE Design Automation Conference (DAC), pp 1–6
12. Putra SD, Ahmad AS, Sutikno S, Kurniawan Y (2018) Attacking AES-masking encryption device with correlation power analysis. IJCNIS 10(2):6
13. Akkar M-L, Giraud C (2001) An implementation of DES and AES, secure against some attacks. In: Koç ÇK, Naccache D, Paar C (eds) Cryptographic Hardware and Embedded Systems — CHES 2001, vol. 2162. Springer, Berlin/Heidelberg, pp 309–318
14. Lu Q, Fan J, Sham C, Lau FCM (2014) A high throughput Gaussian noise generator. In: 2014 IEEE Asia Pacific conference on circuits and systems (APCCAS). IEEE, Piscataway, pp 117–120
15. Kamoun N, Bossuet L, Ghazel A (2009) Correlated power noise generator as a low cost DPA countermeasures to secure hardware AES cipher. In: 2009 3rd International Conference on Signals, Circuits and Systems (SCS), Medenine, Tunisia, pp 1–6
16. Kamoun N, Bossuet L, Ghazel A (2011) A masked correlated power noise generator use as a second order DPA countermeasure to secure hardware AES cipher. In: ICM 2011 Proceeding, Hammamet, Tunisia, pp 1–5
17. Sumari ADW, Ahmad AS, Wuryandari AI, Sembiring J (2012) Brain-inspired knowledge-growing system: towards a true cognitive agent. Int J Comp Sci Artif Intell 2(1):26–36
18. Sumari ADW, Ahmad AS, Wuryandari AI, Sembiring J (2010) Constructing brain-inspired knowledge-growing system: a review and a design concept. In: 2010 international conference on distributed frameworks for multimedia applications. IEEE, Piscataway, pp 1–7
19. Hall DL, McMullen SAH (2004) Mathematical techniques in multi-sensor data fusion, 2nd edn. Artech House, Boston
20. Das SK (2008) High-level data fusion. Artech House, Boston
21. Sumari ADW, Ahmad AS (2008) Design and implementation of multi Agent based information fusion system for decision making support (a case study on military operation). J ICT Res Appl 2(1):42–63–63
22. Bachri KO, Khayam U, Soedjarno BA, Sumari ADW, Ahmad AS (2019) Cognitive artificial-intelligence for Doernenburg dissolved gas analysis interpretation. TELKOMNIKA 17(1):268–274
23. Daemen J, Rijmen V (2002) The design of Rijndael. Springer, Berlin/Heidelberg
24. Putra SD, Sutikno S, Kurniawan Y, Ahmad AS Design of an AES device as device under test in a DPA Attack. Int J Netw Secur 20:10
25. Putra SD, Ma'muri, Sarwono S, Kurniawan Y, Ahmad AS (2018) Design of an AES device as device under test in a DPA attack. Int J Netw Secur 20:256–265
26. Putra SD, Ahmad AS, Sarwono S, Kurniawan Y, Sumari ADW (2018) Revealing AES encryption device key on 328P microcontrollers with differential power analysis. IJECE 8:5144
27. Cormen TH (ed) (2009) Introduction to algorithms, 3rd edn. MIT Press, Cambridge, MA

Dr. Septafiansyah Dwi Putra, S.T., M.T. received the S.T. (bachelor degree) in electrical engineering from Lampung University and Magister Teknik (M.T.) degree in Computer Engineering with Cum Laude, and Doctor (Dr.) in Electrical Engineering and Informatics also with Cum Laude from Institut Teknologi Bandung (ITB). His research interests are cognitive artificial intelligence and cybersecurity. He is a lecturer and researcher at Cybersecurity Research Group, Management of Informatics - Politeknik Negeri Lampung, Lampung, Indonesia. Dr. Septafiansyah also hold several professional certificates in cybersecurity and computer system security.

Colonel (Electronic) Dr. Ir. Arwin Datumaya Wahyudi Sumari, S.T., M.T., IPM, ASEAN Eng. was inaugurated as 2nd Lieutenant Officer of Electronics Corps of the Indonesian Air Force from Indonesian Air Force Academy, Yogyakarta, and achieved Adi Makayasa Medal as Best of the Best Graduate. He received S.T. degree in Artificial Neural Network, M.T. degree in Multi Agent System, and Doctor in Cognitive Artificial Intelligence from Institut Teknologi Bandung (ITB), Indonesia in April 1996, March 2008, and July 2010. All degrees were achieved with Cum Laude. His research interests are Cognitive Artificial Intelligence, multi agent systems, and cybersecurity. He is a Senior Researcher at CAIRG, ITB. He has written more than 200 technical and general papers published internationally and nationally, and is also Steering Committee in several International Conferences. Colonel Dr. Arwin Sumari holds several professional certifications. He currently is a Senior Electrical Engineer Officer at Abdulrachman Saleh Air Force Base, 2nd Operation Command, the Indonesian Air Force at Malang, East Java, Indonesia. He has two university affiliations where he is an Adjunct Professor at State Polytechnic of Malang since 2019 and also an Assistant Professor at Indonesia Defense University since 2012.

Prof. Adang Suwandi Ahmad, Dr.-Ing., DEA, Ir., IPU (ASA) received his engineering degree (Ir.) in Electrical Engineering from ITB in 1976, Diploma Etude Approfondi Signaux et Bruits (DEA) option Electronique, and Docteur Ingenieur Signaux et Bruits option Electronique (Dr.-ing) from Universite des Sciences du Languedoc Montpellier, France in 1978 and July 1980 respectively. He became Institut Teknologi Bandung's Professor in Intelligent Electronics Instrumentation System in 2000. ASA past researchs were in Electronics Instrumentation Systems (Devices and Systems) and Intelligent Electronics Systems/Artificial Intelligence. Cooperation with Navy Research Service (1992) had yielded a War Game Simulator. He founded Intelligent System Research Group (ISRG) ITB in 1993 which then became CAIRG. His research focuses on Cognitive Artificial Intelligence, information sciences, intelligent computations, and multi agent systems. He is the former Dean of the School of Electrical

Engineering and Informatics ITB, Bandung, Indonesia. Prof. Adang Suwandi Ahmad, who initiated Cognitive Artificial Intelligence (CAI), passed away on July 2019 leaving his brilliant legacies to all of his students to carry on his thoughts to the world. Rest in Peace Professor.

Sarwono Sutikno, Dr.Eng., CISA, CISSP, CISM, CSX-F received B.S in Electronics degree from Institute Technology of Bandung, Bandung, Indonesia, in 1984, and received the Master of Engineering degree and Doctor of Engineering degree in Integrated System from Tokyo Institute of Technology, Tokyo, Japan in 1990 and 1994, respectively. His research interests focus on the implementation of cryptographic algorithms in Integrated Circuits including Embedded System Security. His Security Engineering focus includes Information Security Management System. He holds several professional certifications including Certified Information System Auditor and ISMS Provisional Auditor, he is also appointed as ISACA Academic Advocate. Currently, he is an advisor to the Corruption Eradication Commission Republic of Indonesia.

Dr. Yusuf Kurniawan, S.T., M.T. received the B.S. degree, master degree, and doctoral degree in electrical engineering from Institut Teknologi Bandung (ITB), Bandung, Indonesia, in 1994, 1997, 2007, respectively. He is currently as a lecturer of School of Electrical Engineering and Informatics ITB. His research interests focus on the design of block cipher and cryptology.

On the Secure Routing Protocols, Selfishness Mitigation, and Trust in Mobile Ad Hoc Networks

Uttam Ghosh, Pushpita Chatterjee, and Al-Sakib Khan Pathan

Abstract In the era of big data, the data are produced by numerous sources. Though, Mobile Ad hoc Network (MANET) would not be directly related to big data technology, there are at least two issues that relate MANET with big data scenario which are: (i) collecting reliable data securely from MANET (ii) obtaining meaningful data from the huge data sets and transmission of that securely through MANET. This is why it is needed to talk about the secure routing protocols in MANET as in some way; such network setting also would be related to contributing to the big data environment. The intent of this chapter is to present a survey on the secure routing protocols in MANET.

1 Introduction

The tremendous advancement in nomadic communication and wireless hand-held devices yields the communication network paradigm known as Mobile Ad hoc Network (MANET), where hand-held mobile devices are collectively organized in a network without any preexisting infrastructure (see Fig. 1). The devices are commonly referred to as nodes. The main applications of such networks can be found in emergency conditions like earthquakes, Tsunami, and other natural disasters, unmanned terrain explorations, defense related applications, etc. Also, MANET is considered the foundation of Wireless Sensor Network (WSN), Vehicular Ad hoc Networks (VANET), Wireless Mesh Network (WMN), and the pervasive networks.

U. Ghosh (✉)
Vanderbilt University, Nashville, TN, USA
e-mail: uttam.ghosh@vanderbilt.edu

P. Chatterjee
Old Dominion University, Norfolk, VA, USA

A.-S. K. Pathan
Department of Computer Science and Engineering,
Independent University, Bangladesh (IUB), Dhaka, Bangladesh
e-mail: spathan@ieee.org

Z. Md. Fadlullah, A.-S. Khan Pathan (eds.), *Combating Security Challenges in the Age of Big Data*, Advanced Sciences and Technologies for Security Applications, https://doi.org/10.1007/978-3-030-35642-2_5

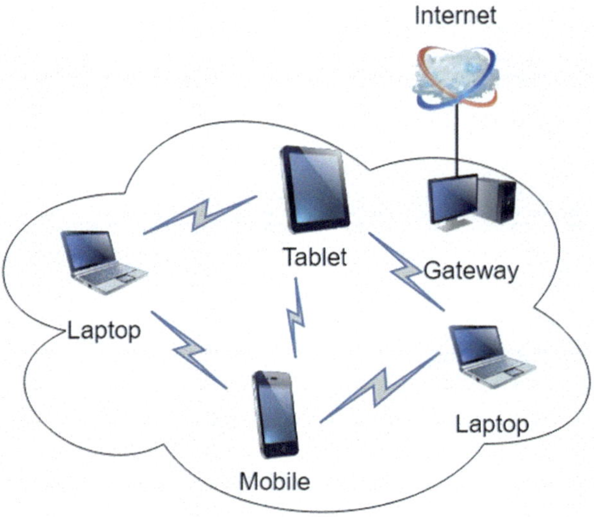

Fig. 1 Mobile ad hoc network

MANETs are useful for commercial and civilian applications like managing hospitals, classrooms, seminar halls, shopping malls, etc., as well.

This kind of network has already gained popularity. However, some inherent features of a MANET give rise to some challenges to the developers for various practical application scenarios. In the last few years, extensive research works have been carried out on the issues like routing, location management, connectivity, security, and other related fields. Security, scalability, robustness, availability are some of the most explored issues in this paradigm.

In this chapter, we present a review of the state-of-the-art research carried out in various fields of MANET like secure routing along with selfishness mitigation and trust aware security solutions. Secure end-to-end delivery is one of the most important issues related to this kind of infrastructure-less distributed networks.

2 Secure Routing in Manet

In a MANET, a node can communicate directly with nodes within its radio range. If a node *S* has to send a packet to *D* which is not in its radio range, it has to rely on other intermediate nodes to forward the packet to reach D. As it is an infrastructure-less dynamic network, wired routing protocols are almost inapplicable. Therefore, specialized routing protocols are required which could adapt dynamically to the changing topologies in MANET. Routing protocols can be broadly classified into three categories based on the underlying routing information update mechanism employed: reactive (on-demand), proactive (table driven), and hybrid.

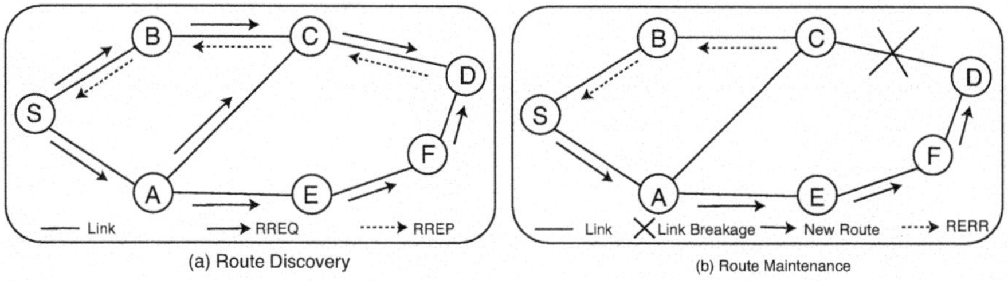

Fig. 2 AODV (**a**) Route Discovery: source node S floods a RREQ message to the entire network; D unicasts RREP message back to S; computed route is S-B-C-D. (**b**) Route Maintenance: C detects link failure to D; C sends RERR message through B to S; S discovers new route S-A-E-F-D

In a reactive routing protocol, it is initiated on a demand basis i.e., whenever source requires path to the destination. There is no need to maintain routing tables between all the nodes at all times. Therefore, it utilizes network energy and bandwidth resources more efficiently, at the cost of increased route discovery delay. Ad-hoc on demand distance vector (AODV) [1] and dynamic source routing (DSR) [2] are the popular protocols in this category.

AODV has route discovery, route maintenance and neighbor maintenance phases. Figure 2a illustrates an AODV route discovery process, where source node S wishes to send data packets to destination node D for which it has no route. S broadcasts a RREQ (Route Request) message which is flooded to all nodes in the network. When D receives multiple RREQ messages from C and F, it computes the shortest path in terms of hop-count from source to destination. Thereafter, D unicasts RREP (Route Reply) message back to S using the reverse path C and B.

In route maintenance phase, each node monitors the link status of the next hops in active routes. When a node detects a link break in an active route, it unicasts a Route Error (RERR) message along the reverse route towards source node. The route maintenance process is shown in Fig. 2b. Here, node C detects the link break and sends RERR to S through B. On receiving RERR message, S re-initiates the route discovery process and discovers a new route S-A-E-F-D again. In neighbor maintenance phase, each node periodically sends HELLO messages to keep the track of its neighboring nodes.

DSR is designed based on the concept of source routing. The source knows the complete hop-by-hop route to the destination. The routes are stored in a route cache. DSR uses route discovery and route maintenance phases. It works in a similar way like AODV except that DSR caches entire route information in each node and does not have neighbor maintenance phase. Figure 3a shows a DSR route discovery process, where source node S wants to communicate with destination node D. As no route exists, S broadcasts a RREQ message which is flooded to all nodes in the network. After receiving RREQ, each intermediate node appends its address and rebroadcasts towards D. On receiving multiple RREQ messages from C and F, D computes the shortest path from source to destination and uses the route cache to unicast a RREP message back to S. The route maintenance phase is shown in Fig. 3b

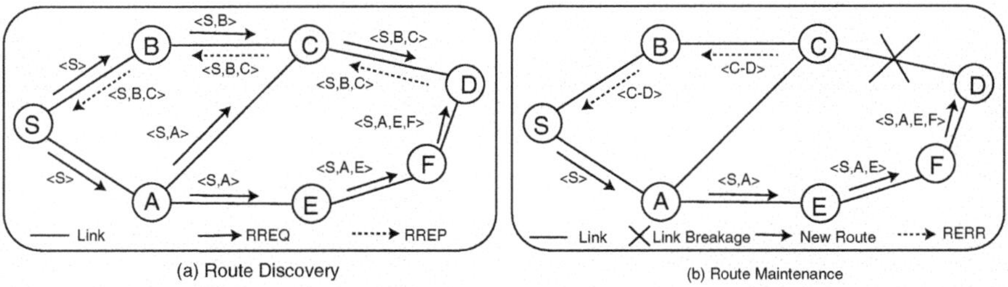

Fig. 3 DSR (**a**) Route Discovery: source node S floods a RREQ message to the entire network; Each intermediate node forwards RREQ after appending its address; D sends RREP message back to S with entire route information <S, B, C > in a cache. (**b**) Route Maintenance: C detects link failure to D; C sends RERR message through B to S; S either uses another route from its cache or it discovers new route <S-A-E-F> to D

where C detects the link failure and sends a RERR to S. On receiving RERR, S removes link and its route cache reinitiates the route discovery process.

In a proactive routing protocol, such as DSDV [3], each node maintains the network topology information in the form of routing tables. These tables are maintained by periodically exchanged routing information, which is generally flooded throughout the network. Whenever a node requires a path to a destination, it runs an appropriate path finding algorithm on the topology information it maintains. If bandwidth and energy resources permit, it is suitable for MANET due to its low route discovery latency.

Hybrid routing protocols like zone routing protocol (ZRP) [4] combine the best features of both reactive and proactive routing protocols. Each node uses proactive routing protocols to reach nodes within certain geographical area (zone), and reactive routing protocols for the rest.

In the following sections, some well-known security schemes will be discussed with the merits and demerits of the same. These are designed to provide security to reactive and proactive routing protocols using both symmetric and asymmetric key cryptography.

2.1 Secure Reactive Routing Protocols

Reactive routing protocols such as DSR and AODV assume that participating nodes do not maliciously disrupt the operation of the protocol. Secure routing protocols cope with malicious activities like modification of routing information, fabricating false routing information, impersonation, etc. These protocols are either completely new stand-alone protocols, or in some cases incorporations of security mechanisms into existing protocols. Routing protocols incorporate conventional authentication and encryption schemes based on cryptography to provide a first line of defense. These include asymmetric and symmetric cryptography. Cryptographic primitives such as hash functions (message digests) can be used to enhance data integrity in transmission as well. Threshold cryptography can be used to hide data by dividing it into a number of shares. Digital signatures can be used to achieve data integrity and authentication services as well. As a second line of defense, trust/reputation mechanisms are implemented with the routing protocols in MANET to defend against attacks or enforce cooperation, reducing selfish node behavior.

2.1.1 Secure Routing Based on DSR

Ariadne [5] is a secure on-demand ad hoc routing protocol based on DSR. It prevents attackers or compromised nodes from tampering with routes consisting of legitimate nodes, and many types of Denial-of-Service (DoS) attacks. It uses only highly efficient symmetric cryptographic primitives. One-way hash functions are used to verify if any hop has been omitted on the route, which is known as per-hop hashing. Ariadne can authenticate messages one of three ways: sharing secrets between each pair of nodes, sharing secrets between communicating nodes combined with broadcast authentication, or digital signature. Ariadne uses pair-wise shared keys, avoids the need for time synchronization but at the cost of higher key setup overhead.

Papadimitratos et al., proposed the secure routing protocol (SRP) [6] that can be used with DSR and ZRP. SRP provides end-to-end authentication with the addition of several security extensions. SRP can detect modification of the route request (RREQ) at the target and route reply (RREP) at the source. However, it does not attempt to prevent unauthorized modification of fields that are ordinarily modified in the course of forwarding packets. A shared secret is established between two nodes, and the non-mutable fields of the exchanged routing messages are protected by this shared secret. The scheme is robust in the presence of a number of non-colluding malicious nodes, and provides accurate routing information in a timely manner. SRP makes no assumption regarding the intermediate nodes, which exhibits arbitrary and malicious behavior. Nodes use secure message transmission (SMT) [7] to ensure secure successful delivery of data packets.

2.1.2 Secure Routing Based on AODV

Sanzgiri et al., develop authenticated routing for ad hoc networks (ARAN) [8], which is based on AODV and utilizes cryptographic public key certificates signed by a trusted third party. The certificate associates an IP address with a public key in order to achieve the security goals of authentication, message integrity and non-repudiation to the route discovery process. The cost of ARAN is a larger routing packet, which results in a higher routing load and latency in route discovery due to the cryptographic computation.

Zapata et al., proposed secure AODV (SAODV) [9], to enforce security in AODV. The idea behind SAODV is to use a signature to authenticate most fields of RREQ and RREP. Two mechanisms are used to secure the AODV messages: digital signature to authenticate the non-mutable fields of the message and hash chain to secure the hop count information [10, 11]. Nodes authenticate AODV routing packets with an SAODV signature extension that prevents certain impersonation attacks. Since the protocol uses asymmetric key cryptography for digital signature, it requires the existence of a key management mechanism that enables a node to acquire and verify the public key of other nodes in the network.

Ghosh et al., proposed an identity-based scheme, Secure Dynamic Routing Protocol (SDRP) [12] which uses digital signature and message authentication code algorithms to provide end-to-end, hop-to-hop and whole-route authentications. The protocol has several advantages over the existing RSA-based secure routing solutions as it requires fewer signature generations and verifications on a route. SDRP uses an identity-based scheme with a small-size of public key, which is certificateless, thus saving routing overhead (RO) and storage. ID-based scheme secures AODV and transmits TCP data to the authorized hosts. The authors also proposed a RSA based Scheme [13] that uses a SAS (Sequential Aggregate Signatures) to secure AODV. The scheme can securely generate the session key for the MANET nodes to secure the TCP.

2.2 Secure Proactive Routing Protocols

Hu et al., proposed secure efficient ad hoc distance vector routing protocol (SEAD) [14] based on DSDV. It is robust against multiple uncoordinated attackers creating incorrect routing state at other nodes in the network. Efficient one-way hash chains are used in the authentication of the sequence number and the metric (hop count) field of a routing table update message.

Zhao et al.'s work presented in [15] is a secure routing protocol which uses identity-based cryptography in a proactive security approach. The authors name the protocol "*proactive*" or "*preclusive*" because they assume security at the beginning of starting the operation, and preclude insecurity proactively. They show the comparative advantage against some other alternative secure routing protocols in terms of routing setup and maintenance. As the protocol does not need any side channel or secret channel at all, that simplifies the lower layer design and saves administrative overhead. Also, it does not use flooding to set up initial routing and does not use multicast to update secret, which improves efficiency. However, this protocol has a tricky and strong assumption that the authors are always able to update system secret before the adversary nodes in surrounding area can compromise a number of nodes and break the secret. In some controlled deployment scenario, this may be achieved however, for many cases, it would be impractical. Even the authors also mentioned about this assumption, "*This is an essential bedrock of the security of the system, but is the most tricky assumption.*" And "*If we can achieve this, we can exclude the adversary nodes all through.*" Hence, implementation of this protocol in most of the practical/usual cases would be really difficult.

The secure link-state protocol (SLSP) [16] is proposed by Papadimitratos et al., which provides a proactive secure link state routing solution for ad hoc networks. It uses digital signature and one-way hash chain to ensure the security of link-state updates. SLSP can be used as the intra-zone routing protocol in ZRP. It is a periodic protocol that receives link-state information through a periodic neighbor location protocol (NLP). When receiving a link state update (LSU) packet, nodes verify the attached signature using a public key that they have cached in the pubic key distribution phase of the protocol and authenticate the hop count by one-way hash chains. Link state information was broadcasted periodically using NLP to detect discrepancies between IP and MAC addresses. SLSP offers protection against individual malicious node by securing neighbor discovery process. However, SLSP is vulnerable to colluding attackers that fabricate non-existing links between themselves and flood information to colluding neighbors.

From the above discussion, it becomes clear that security protocols mainly focus on authentication based on cryptographic techniques as a first line of defense. This may prevent a MANET from being attacked by outsider malicious nodes. However, an authenticated node can still compromise the MANET or may behave maliciously or selfishly. Therefore, authentication based techniques are not sufficient to prevent insider attacks. We need second line of defense to prevent these kinds of attacks.

Another important category of secure routing is cooperation enforcement between the nodes; thus increasing the availability. If the primary goal is to increase the availability and overall throughput, and achieve the robustness of the network, the cooperation enforcement techniques may fit better. In the next section, a brief review of works done on such techniques is discussed.

3 Selfishness Mitigation

Cooperation enforcement approaches are categorized as reputation based (based on reputation building, supports monitoring of neighbors' activities) and credit based (based on economic incentives: pricing requires the existence of tamper-resistant hardware or a virtual bank). These approaches can be used in collaboration with the existing secure routing protocols to provide comprehensive security for the data in the network.

3.1 Reputation Based Schemes

The reputation based schemes use the reputation of the nodes to forward packets through the most reliable nodes. The reputation value (RV) of a node is measured by its behavior towards forwarding others' packets. RV increases if the node rightly forwards the packets of its neighbors without modifying them, and decreases otherwise. They also incorporate techniques to isolate the misbehaving nodes (nodes with a low RV). Depending on the type of observation about a neighbor node, the reputation based models can be further divided into two subclasses: models where the reputation is based only on a node's personal/self observation (first-hand reputation information) and models according to which recommendations/observations of other nodes are taken into consideration (second-hand reputation exchanges). If a node observes that another node does not behave rightly, it reports this observation to the other nodes in the MANET.

Buchegger et al., design a protocol namely *Cooperation Of Nodes: Fairness In Dynamic Ad-hoc NeTworks* (CONFIDANT) [17] as an extension to DSR. The scheme facilitates monitoring and reporting for a route establishment that avoids the misbehaving nodes. Both direct and indirect observations are used to detect a misbehaving node. Revocation and reintegration of a non-malicious node into network is permissible in CONFIDANT if the node is incorrectly accused or turns out to be a repentant and no longer malicious. The disadvantage here is the requirement of a pre-existed trust relationship.

The first version of CONFIDANT was vulnerable to rumor spreading phenomena [18]. Further, this problem has been addressed through a Bayesian model [19, 20] that classifies and excludes the liars. Both positive and negative reputations are used to calculate a cooperation factor that consists of the frequency of misbehavior in relation to the cumulative activity of the node.

CORE [21], introduced by Michiardi et al., relies on the DSR routing protocol. It uses first and second-hand experiences, combined by a specialized function, which is used by the Watchdog mechanism [22]. The CORE scheme is immune to attacks; as no negative ratings are spread; the malicious decrease of node's reputation is not possible. CORE gradually isolates misbehaving nodes when the reputation assigned to a neighboring node falls below a predefined threshold. However, misbehaving nodes can be reintegrated into the network if they purposefully increase their reputation by cooperating with the network operation. CORE does not discriminate between malfunctioning and misbehaving nodes. It assumes that every node uses identical calculations of the RV, assigning the same weights to the same functions. As MANETs consist of devices equipped with different resources providing discrete services, they prefer to use different levels of importance on functions [23].

Bansal et al., proposed the Observation-based Cooperation Enforcement in Ad hoc Networks (OCEAN) [24], based on DSR, which introduces an intermediate layer between the network and the Medium Access Control (MAC) layers for making intelligent routing decisions. Every node maintains an *avoid list* through the ratings for each neighboring node. Checking this list, a route is rated good or bad, and eventually misbehaving nodes are isolated. However, a second chance mechanism is used to allow nodes that misbehaved in the past to become operational again. OCEAN uses a credit-based policy, to deal with nodes that do not participate in the route discovery process. It does not require any tamper-proof hardware or a centralized server. However, as OCEAN is sensitive to the tuning of the faulty threshold parameter, second-hand schemes perform better over a broader range of tunings. Additionally, it is not effective in reducing the throughput of misbehaving nodes and takes no countermeasures to prevent collusion.

The Secure and Objective Reputation-based Incentive (SORI) Scheme [25], proposed by He et al., focuses on the packet forwarding function. SORI combines feature of the first-hand and reputation spreading schemes. It takes into account the credibility of the nodes which contribute to the calculation of the reputation. This makes it difficult for an attacker to test multiple identities, trying to impersonate one identity in order to improve its reputation. This mechanism is designed to treat generously the nodes that do not intentionally drop packets. The security mechanism is based on a one-way hash chain and MAC. SORI takes no countermeasures to prevent collusion.

Dewan et al., introduce a first-hand reputation information model [26], based on AODV. It uses acknowledgements to observe the behavior of adjacent nodes, rather than complex operations to decide the reputation of a node. The source node finds a set of paths to a destination using AODV. The first hop node forwards packet to the next hop node with the highest reputation and the process is repeated till the packet reaches its destination. The destination acknowledges the packet to the source. The corresponding entry of the reputation table is updated by rewarding the first hop. If a non-cooperative node resides in the route and drops packet, source may not receive an acknowledgment within a predefined interval. A load balancing method that balances the load among the well-reputed nodes might overcome such phenomena. It does not include an explicit mechanism for giving a second chance to nodes that experience relay failures or have low recourses. However, the authors proposed two techniques that extend the basic scheme and handle these situations.

3.2 *Credit Based Schemes*

For credit based models, the packet forwarding task is treated as a service which can be evaluated and charged. These models incorporate a form of virtual currency to regulate the dealings between the various nodes for packet forwarding. They require the existence of tamper-resistant hardware or a virtual bank.

Tamper resistance is basically some kind of resistance to tampering or intentional sabotage by either the normal users of a system or others with physical access to the device/hardware. On the other hand, virtual bank offers trusted third party services to the nodes. Sprite, a simple, cheat-proof, credit-based system for MANETs [27] was proposed by Zhong et al. It does not require tamper-proof hardware but incorporates a centralized credit clearance service (CCS). The CCS believes that a node has forwarded a packet if there is a successor of that node on the path reporting a valid receipt of the packet. A potential disadvantage of Sprite is the assumption that a fast connection to the CCS is needed for reporting the obtained receipts. A generalization of Sprite that encourages the participation of nodes during the route discovery is also introduced.

Another Scheme [28], introduced by Yang et al., protects both routing and packet forwarding in the context of AODV protocol. It is self-organized and does not assume existence of any a-priori trust between the nodes or centralized trust entity. It isolates the misbehaving nodes and employs threshold cryptography to enhance the tolerance against these nodes. Nodes actively and collaboratively monitor others' traffic to detect any misbehavior. The neighbor verification employs the RSA based cryptographic primitives. Regarding the key setup complexity and the requirement for the threshold cryptography, the authors mention that; when light-weight cryptography is employed, the computation complexity is decreased whilst hashing techniques might decrease the storage overhead.

Anderegg et al. propose the ad hoc-VCG Scheme [29] based on DSR. It is a credit-based model based on a second best sealed type of auction. The adhoc-VCG scheme estimates this cost through the cost-of-energy parameter. If an intermediate node does not get a payment to cover its forwarding costs, it refuses to forward. The nodes determine the energy emission levels to reach their neighbors using a signaling process and additional control fields on packets. The ad hoc-VCG may fail in the presence of collusion of nodes, trying to maximize their payments. Moreover, it requires complete knowledge of the network topology to construct the graph, which creates significant overhead during the route discovery phase. Finally, it does not focus on the actual payment delivery, but only on the estimation of the payments.

The protocols that have been discussed so far provide security to routing protocols either by using cryptography primitives or using trust and reputation based schemes to ensure security and availability. Apart from those, there are few protocols which rely on trusted framework to achieve security objective. In the following section, some other trust based protocols are briefly described.

4 Trust Management Schemes

This section summarizes the trust management schemes that have been developed for ad hoc networks. We describe trust management schemes based on specific design purposes such as secure routing, authentication and key management. Further, we also describe the existing general frameworks for trust (or reputation) evidence distribution and evaluation.

4.1 Secure Routing Using Trust

Most reputation-based trust management schemes are devised for collaborative secure routing by detecting misbehaving nodes that are either selfish or malicious. The cooperation enforcement schemes for selfish nodes are described in Sect. 3 and different secure routing schemes have been already discussed in Sect. 2. In some of the secure routing protocols, a-priori trust relationships are assumed. Here, some other trust based secure routing protocols are discussed.

Nekkanti et al., proposed an extension to AODV [30] using trust factor and security level at each node. This approach deals differently with each RREQ based on the node's trust factor and security level. The routing information for every request is encrypted which leads to large overheads. However, the approach does not address evaluation of trust itself.

Pirzada et al. proposed and examined the efficacy of trust based reactive routing protocols in the presence of attacks. This work [31] only considers first hand information to evaluate others' trust values. Their trust evaluation scheme is restricted to direct neighboring nodes. Pissinou et al. devised a secure AODV based routing protocol [32] for multi-hop MANET to discover a secure end-to-end route. The protocol calculates the trust values based only on direct observations, assuming that trust is transitive. Ghosh et al., enhanced trust management in their proposal [33] by considering the confidence level of trust. The confidence level is used as a weight on the computed trust value and the method for calculating trust in a fully distributed way provides a general framework that can be applied to non-trust aware routing protocols.

Zouridaki et al., proposed a trust establishment mechanism [34] called *Hermes* to improve the reliability of packet forwarding over multi-hop routes in the presence of potentially malicious nodes. Direct observations are used to evaluate opinions about others. As an extension, Zouridaki et al., employed both first-hand trust information and second-hand trust information forwarded from neighboring nodes about non-neighboring nodes. This trust establishment Scheme [35] can cope with more attacks, including propagation of false recommendations or information, identifying bad nodes among neighboring nodes, colluding attacks, replay attacks, and duplicate attacks.

Li et al., also extended AODV and adopted a trust model to guard against malicious behaviors of nodes at the network layer [36]. They represented trust as opinions stemming from subjective logic. They proposed an Objective Trust Management Framework (OTMF) based on both direct and indirect information for reputation management and showed the effectiveness of OTMF. However, this work did not consider node collusion in obtaining second-hand information, which may lead to incorrect recommendations. Sun et al., proposed trust modeling and evaluation methods [37] for secure ad hoc routing and malicious node detection. The unique part of their design is to consider trust as a measure of uncertainty that can be calculated using entropy. However, this work considers packet dropping as the only component of direct observations to evaluate trust.

Abusalah et al., proposed a trust aware routing protocol (TARP) [38] and developed a trust metric based on six trust components including software configuration, hardware configuration, battery power, credit history, exposure and organizational hierarchy. However, no consideration was given to trust decay over time and space to reflect uncertainty due to dynamics and incomplete information in MANET environments.

Balakrishnnan et al. developed a trust model [39] to strengthen the security and to deal with the issues associated with recommendations. This work uniquely considered a context dependency characteristic of trust in extended DSR. Their model utilizes only trusted routes for communication, and isolates malicious nodes based on the evidence obtained from direct interactions and recommendations. In quorum or threshold cryptography schemes, a node must successfully interact with at least k of n distributed trusted authority (TA) nodes. Finding k such nodes can be resource intensive. Reidt et al., prioritize the TA nodes in their proposal [40]. They find a route to connect to *k* desirable TA nodes so as to minimize a performance metric such as overhead, taking into account reliability, and energy consumption of individual nodes.

Wang et al. proposed a mechanism [41] to distinguish selfish peers from cooperative ones based solely on local observations of AODV. They use a finite state machine model of locally observed AODV actions to construct a statistical description of each peer's behavior. A series of well-known statistical tests are applied to features obtained from the observed AODV actions.

4.2 Authentication Using Trust

There have been efforts to establish trust relationships to ensure authentication in ad hoc networks. Weimerskirch et al., developed a trust model [42] based on human behavior, noting that society can be considered as an ad hoc network. They used recommendations from a distributed trust model to construct trust relationships and extended the proposal by adding a request for recommendations. The assumption of low value transactions does not require any evidence based mechanism to ensure trust such as authentications using public/private keys. Consequently, it is not applicable to systems where hostility may be high, or where consequences of misplaced trust can be severe.

Verma et al., presented an overview of a trust negotiation Scheme [43] using DSR and ZRP. This scheme consists of two components. The peer-to-peer component deals with secure communications with neighbors in a lightweight manner. The main goal of this work is to add robustness to the process of trust negotiation rather than trust evaluation. Pirzada et al. proposed a trust based communication model [44] based on a notion of a belief. It provides a dynamic measure of reliability and trustworthiness in MANETs. The merit of this work can be precisely identified as it incorporates utility as general trust and time as situational trust into the overall trust metric to evaluate an agent in the network. However, the situational trust considered is limited to monitoring dynamics of packet forwarding behaviors.

Ngai et al. proposed a secure public key authentication service using a trust model [45] to prevent propagation of false public keys in the presence of malicious nodes. Trust is evaluated based on direct monitoring as well as recommendation. However, this work does not consider group membership changes, the distance from the evaluator, and their effect on the performance of the trust management scheme.

4.3 Key Management Using Trust

A survey of key management techniques for network layer security is presented in [46]. Virendra et al. proposed a trust based security architecture [47] for key management in MANET. It aims to establish keys between nodes based on their trust relationships and to build a secure distributed control framework using trust as a metric. The unique part of this work is that it considers the trust level of each node in a physical as well as a logical sense. However, establishing pairwise keys based on pairwise trust relation may not be feasible in terms of scalability and in the presence of high network dynamics in a large network.

Li et al. demonstrated an on-demand, fully localized, and hop-by-hop public key management protocol [48] for MANETs. In this protocol, each node generates its own public/private key pair, issues its certificate to neighboring nodes, and provides authentication service by adapting to the dynamic network topology, without reliance on any centralized server. However, only certificate chains are used to derive trust.

Chang et al., proposed a Markov chain trust model [49] to obtain the trust values (TVs) for one-hop neighbors. They designed a trust based hierarchical key management scheme by selecting a certificate authority server (CA) and a backup CA with the highest TVs. This work gives a rigorous analysis of TVs and considers a variety of attacks.

4.4 Trust Evidence Distribution and Evaluation

Several trust management schemes have been proposed in order to provide a general framework for trust evidence distribution or evaluation in ad hoc networks. Yan et al. proposed a trust evaluation mechanism based security solution [50] for data protection, secure routing and other network activities. This trust evaluation model is called *personal trusted bubble* (PTB). It considers many factors including experience statistics, data value (the higher the value of the data, the higher is the trust needed from other PTBs to transfer it), intrusion black list, reference, personal preference, and PTB policy.

Jiang et al. proposed Ant Based trust Evidence Distribution (ABED) [51] based on the swarm intelligence paradigm, which is highly distributed and adaptive to mobility. In ABED, pheromones are deposited at nodes by mobile agents called ants and pheromones provide the mechanism for information exchange and interactions. However, no specific attackers are considered to prove the robustness of the scheme in presence of attacks. In the continuing work, Jiang et al. [52] addressed distributed trust computation and establishment using random graph theory and the theory of dynamic cooperative games. Trust relationships are ternary (yes, no, don't care) and the emphasis is on understanding steady state behaviors. This model incorporates trust variables with continuous value, dynamics, and transient behaviors.

Theodorakopoulos et al. proposed a trust evidence evaluation Scheme [53] for MANETs. The evaluation process is modeled as a path problem in a directed graph where vertices represent entities and edges represent trust relations. Their case study uses the Pretty Good Privacy (PGP) web of trust to express an example of trust model based on semirings and shows that their scheme is robust in presence of attackers. However, their work assumes that trust is transitive. Further, trust and confidence values are represented as binary rather than continuous values. Even though no centralized trusted third party exists, their work makes use of a source node as a trusted infrastructure, which incurs vulnerability in MANET.

Boukerche et al. proposed a distributed reputation management mechanism [54] known as generalized reputation evaluation (GRE), using a comprehensive computational reputation model. GRE seeks to prevent malicious nodes from entering a trusted community. However, no specific attack model was addressed.

Cho et al., proposed a trust management Scheme [55] for group communication systems in MANETs. This work proposed a composite trust metric reflecting various aspects of a node such as sociability (i.e., social trust) and task performance capability (i.e., QoS trust), and investigated the effect of the trust chain length used by a node to establish acceptable trust levels through subjective trust evaluation.

Chatterjee et al., proposed a distributed secure trust aware clustering protocol [56] that provides secure solution for data delivery. The proposed trust model calculates the trust of a node using self and recommendation evidences of its one-hop neighbors. The proposed clustering protocol organizes the network into one-hop disjoint clusters and elects the most qualified, trustworthy node as a cluster-head. The cluster-head election is made secure by an authenticated voting scheme that uses parallel multiple signatures.

5 Conclusions

In our investigation of the area, we have found that most of the secure routing protocols are based on complex cryptographic computation or key management using trusted third party for key distribution. Mitigating selfishness of nodes in MANET is an important issue to be handled to achieve proper functionality and availability of nodes in the network. Moreover, the security measures must be energy efficient to increase the lifetime of nodes as well as the network. Therefore, though trust and security are often considered separately [11], trust can play a vital role for securing ad hoc routing protocols for MANET (and WSN as well).

References

1. Perkins C, Royer EB, Das S (2003) Ad hoc on demand distance vector (aodv) routing. IETF RFC 3561
2. Hu YC, Perrig A, Johnson D (2004) The dynamic source routing protocol for mobile ad hoc networks (dsr), draft-ietf-manet-dsr-10.txt
3. Perkins EC, Bhagwat P (1994) Highly dynamic destination-sequenced distance-vector routing (dsdv) for mobile computers. Proc ACM SIGCOMM 24:234–244
4. Haas Z, Pearlman M (1998) The performance of query control schemes for the zone routing protocol. Proc ACM SIGCOMM:167–177
5. Hu YC, Perrig A, JohnsonDB (2002) Ariadne: a secure on-demand routing protocol for ad hoc networks. In: Proceedings of the 8th ACM MOBICOM 2002, September 2002
6. Papadimitratos P, Haas ZJ (2002) Secure routing for mobile ad hoc networks. In: Proceedings of SCS communication networks and distributed systems modeling and simulation conference (CNDS 2002), pp 1–12

7. Papadimitratos P, Haas Z (2003) Secure message transmission in mobile ad hoc networks. Elsevier Ad Hoc Netw J 1:193–209
8. Sanzgiri K, Dahill B, Levine B, Shields C, Royer EB (2002) A secure routing protocol for ad hoc networks. In: Proceedings of 10th IEEE ICNP 2002, pp 78–87
9. Zapata MG, Asokan N (2002) Securing ad hoc routing protocols. In: Proceedings of the ACM WiSe, pp 1–10
10. A.-S. K. Pathan and C. S. Hong, "SERP: Secure Energy-efficient Routing Protocol for Densely Deployed Wireless Sensor Networks," Annals of Telecommunications, https://doi.org/10.1007/s12243-008-0042-5, 63, Numbers 9–10, October 2008, pp. 529–541
11. Pathan A-SK (2014, Inderscience Publishers) On the Boundaries of Trust and Security in Computing and Communications Systems. Int J Trust Manag Comput Commun 2(1):1–6
12. Ghosh U, Datta R (2014) Sdrp: secure and dynamic routing protocol for mobile adhoc networks. IET Netw 3(3):235–243
13. Ghosh U, Datta R (2011) Identity based secure aodv and tcp for mobile ad hoc networks. In: Proceedings of the 1st international conference on wireless Technologies for Humanitarian Relief, ACM, pp 339–346
14. Hu Y-C, Perrig A, Johnson DB (2002) Sead: secure efficient distance vector routing for mobile wireless ad hoc networks. In: Proceedings of the 4th IEEE WMCSA 2002, pp 3–13
15. Zhao S, Aggarwal A, Liu S, Wu H (2008) A secure routing protocol in proactive security approach for mobile Ad-Hoc networks. In: 2008 IEEE wireless communications and networking conference, 31 March–3 April 2008
16. Papadimitratos P, Haas Z (2003) Secure link state routing for mobile ad hoc networks. In: Proceedings of the IEEE workshop on security and Assurance in Ad Hoc networks, pp 27–31
17. Buchegger S, Boudec J (2002) Performance analysis of the confidant protocol cooperation of nodes fairness in dynamic ad-hoc networks. In: Proceedings of ACM MOBIHOC 2002, pp 226–236
18. Buchegger S, Boudec JYL (2003) The effect of rumour spreading in reputation systems for mobile ad-hoc networks. In: Proceedings of WiOpt03, March 2003
19. Buchegger S, Boudec JYL (2003) Coping with false accusations in misbehavior reputation systems for mobile ad-hoc networks. Technical report IC/2003/31, Ecole Polytechnique Federale de Lausanne, May 2003
20. Buchegger S, Boudec J (2004) A robust reputation system for p2p and mobile ad-hoc networks. In: Second workshop on the economics of peer-to-peer systems, June 2004
21. Michiardi P, Molva R (2002) Core: a collaborative reputation mechanism to enforce node cooperation in mobile ad hoc networks. In: Proceedings of CMS 2002, pp 107–121
22. Marti S, Giuli TJ, Lai K, Baker M (2000) Mitigating routing misbehavior in mobile ad hoc networks. In: Proceedings of 6th ACM MOBICOM 2000, pp 255–265
23. Marias G, Georgiadis P, Flitzanis D, Mandalas K (2006) Cooperation enforcement schemes for manets: a survey. Wiley's J Wirel Commun Mobile Comput 6:319–332
24. Bansal S, Baker M (2003) Observation-based cooperation enforcement in ad-hoc networks. Techical Report, Stanford University
25. He Q, Wu D, Khosla P (2004) Sori: a secure and objective reputation based incentive scheme for ad-hoc networks. In: Proceedings of IEEE WCNC 2004, vol. 2, pp 825–830
26. Dewan P, Dasgupta P, Bhattacharya A (2004) On using reputations in ad hoc networks to counter malicious nodes. In: Proceedings of tenth international conference on parallel and distributed systems ICPADS 2004, pp 665–672
27. Zhong S, Chen J, Yang YR (2003) Sprite: a simple, cheat-proof, credit-based system for mobile ad-hoc networks. In: Proceedings of IEEE INFOCOM 2003, pp 1987–1997
28. Y. H, M. X, and L. S (2002) Self-organized network-layer security in mobile ad hoc networks. In: Proceedings of ACM WiSe 2002, pp 11–20, September 2002
29. Anderegg L, Eidenbenz S (2003) Ad hoc-vcg: a truthful and cost efficient routing protocol for mobile ad hoc networks with selfish agents. In: Proceedings of MOBICOM 2003, pp. 245–259
30. Nekkanti RK, Lee C (2004) Trust-based adaptive on demand ad hoc routing protocol. In: Proceedings of 42th annual ACM southeast regional conference, pp 88–93

31. Pirzada AA, McDonald C (2004) Establishing trust in pure ad-hoc networks. In: Proceedings of 27th conference on Australasian computer science CRPIT 2004, pp 47–54
32. Ghosh T, Pissinou N, Makki K (2004) Collaborative trust-based routing in multi-hop ad hoc networks. In: Proceedings of 3rd International IFIPTC06 Networking Conference, pp 1446–1451, Lecture Notes in Computer Science, May 2004
33. Ghosh T, Pissinou N, Makki K (2005) Towards designing a trust routing solution in mobile ad hoc networks. Mobile Netw Appl 10:985–995
34. Zouridaki BL Mark MH, Thomas RK (2005) A quantitative trust establishment framework for reliable data packet delivery in MANETs. In: Proceedings of 3rd ACM workshop on security for Ad Hoc and sensor networks, pp 1–10
35. Zouridaki BL, Mark MH, Thomas RK (2006) Robust cooperative trust establishment for manets. In: Proceedings of 4th ACM workshop on security of Ad Hoc and sensor networks, pp 23–34
36. Ruidong L, Jie L, Peng L, Chen HH (2007) An objective trust management framework for mobile ad hoc networks. In: Proceedings of IEEE 65th vehicular technology conference 2007, pp 56–60
37. Sun YL, Yu W, Han Z, Liu KJR (2006) Information theoretic framework of trust modeling and evaluation for ad hoc networks. IEEE JSAC 24:305–317
38. Abusalah L, Khokhar A, Guizani M (2008) A survey of secure mobile ad hoc routing protocols. IEEE Commun Surv Tutorials 19:78–93
39. Balakrishnnan V, Varadharajan V, Tupakula UK, Lucs P (2007) Trust and recommendations in mobile ad hoc networks. In: Proceedings of 10th IEEE international conference on networking and services, pp 64–69
40. Reidt S, Wolthusen SD, Balfe S (2009) Robust and efficient communication overlays for trust authority computations. In: Proceedings of 32nd IEEE Sarnoff symposium 2009, pp 88–92
41. Wang X, Liu L, Su J (2010) Rlm: a general model for trust representation and aggregation. IEEE Trans Serv Comput 5:131–143
42. Weimerskirch A, Thonet G (2001) A distributed light-weight authentication model for ad-hoc networks. In: Proceedings of 4th international conference on information security and cryptology, pp 77–82
43. Verma RRS, O'Mahony D, Tewari H (2001) Ntm – progressive trust negotiation in ad hoc networks. In: Proceedings of 1st joint IEI/IEE symposium on telecommunications Systems Research
44. Pirzada AA, McDonald C, Datta A (2006) Performance comparison of trust-based reactive routing protocols. IEEE Trans Mobile Comput 5:695–710
45. Ngai ECH, Lyu MR (2004) Trust and clustering-based authentication services in mobile ad hoc network. In: Proceedings of 24th IEEE ICDCS workshops, pp 582–587
46. Hegland A, Winjum E, Mjolsnes SF, Rong C, Kure O, Spilling P (2006) A survey of key management in ad hoc networks. IEEE Commun Surv Tutorials 8:48–66
47. Virendra M, Jadliwala M, Chandrasekaran M, Upadhyaya S (2005) Quantifying trust in mobile ad hoc networks. Proc IEEE KIMAS 2005:65–71
48. Li R, Li J, Liu P, Chen HH (2006) On demand public key management for mobile ad hoc networks. Wiley WCMC 6:295–306
49. Chang BJ, Kuo SL (2009) Markov chain trust model for trust value analysis and key management in distributed multicast manets. IEEE Trans Veh Technol 58:1846–1863
50. Yan Z, Prehofer C, (2010) Autonomic trust management for a component based software system. In: IEEE transactions on dependable and secure computing, accepted on April 2010
51. Jiang T, Baras JS (2004) Ant-based adaptive trust evidence distribution in manet. In: Proceedings of 2nd international conference on mobile distributed computing systems workshops, pp 588–593
52. Jiang T, Baras JS (2004) Cooperative games, phase transition on graphs and distributed trust in manets. In: Proceedings of 43th IEEE conference on decision and control, pp. 93–98
53. Theodorakopoulos G, Baras JS (2006) On trust models and trust evaluation metrics for ad hoc networks. IEEE JSAC 24:318–328

54. Boukerche A, Ren Y (2008) A security management scheme using a novel computational reputation model for wireless and mobile ad hoc networks. In: Proceedings of international workshop on modeling analysis and simulation of wireless and Mobile systems, pp 88–95
55. Cho JH, Swami A, Chen IR (2009) Modeling and analysis of trust management for cognitive mission-driven group communication systems in mobile ad hoc networks. In: Proceedings international conference on computational science and engineering, pp 641–650, Springer LNCS, August 29–31 2009
56. Chatterjee P, Ghosh U, Sengupta I, Ghosh SK (2014) A trust enhanced secure clustering framework for wireless ad hoc networks. Springer Wirel Netw 20(7):1669–1684

Uttam Ghosh is an Assistant Professor of the Practice in the Department of Electrical Engineering and Computer Science, Vanderbilt University, Nashville, TN, USA. Dr. Ghosh obtained his PhD in Electronics and Electrical Engineering from the Indian Institute of Technology Kharagpur, India, in 2013, and has postdoctoral experience at the University of Illinois in Urbana-Champaign, Fordham University, and Tennessee State University. He has been awarded the 2018–2019 Junior Faculty Teaching Fellow (JFTF) and has been promoted to a Graduate Faculty position at Vanderbilt University. Dr. Ghosh has published 50 papers and book chapters at books and reputed international journals including IEEE Transaction, Elsevier, Springer, IET, Wiley, InderScience, and IETE, and also in top international conferences sponsored by IEEE, ACM, and Springer. Dr. Ghosh has conducted several sessions and workshops related to Cyber-physical Systems (CPS), SDN, IoT, and smart cities as co-chair at top international conferences including IEEE SECON, CPSCOM, IEMCON, ICDCS, and so on. He has served as a Technical Program Committee (TPC) member at renowned international conferences including ACM SIGCSE, IEEE LCN, IEMCON, STPSA, SCS SpringSim, IEEE Compsac, and many more. He is serving as an Associate Editor of the *International Journal of Computers and Applications*, Taylor & Francis, and is also a reviewer for international journals including IEEE Transactions, Elsevier, Springer, and Wiley. Dr. Ghosh is contributing as Guest Editor for special issues with ACM *Transactions on Internet Technology* (TOIT), Springer *MTAP*, and Wiley *ITL*. He is a Senior Member of the IEEE and a member of AAAS, ASEE, ACM, and Sigma Xi. His main research interests include cybersecurity, computer networks, wireless networks, information centric networking, and software-defined networking.

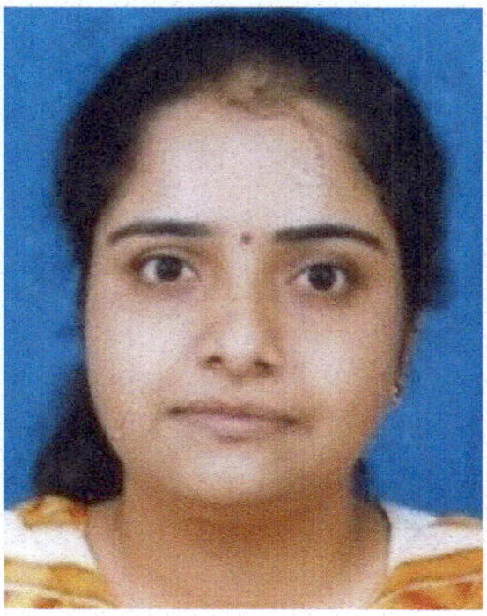

Pushpita Chatterjee is a Research Consultant at Old Dominion University, VA. She received her PhD from Indian Institute of Technology Kharagpur, India, in 2012. Pushpita has a number of publications to her credit in international journals, conferences, and book chapters. Her research interests include mobile computing, distributed and trust computing, wireless ad hoc and sensor networks, information-centric networking, and software-defined networking. She is a member of IEEE.

Al-Sakib Khan Pathan Pathan is a Professor of Computer Science and Engineering. Currently, he is with the Independent University, Bangladesh, as an Adjunct Professor. He received PhD in Computer Engineering in 2009 from Kyung Hee University, South Korea, and BSc in Computer Science and Information Technology from Islamic University of Technology (IUT), Bangladesh, in 2003. In his academic career so far, he worked as a faculty member at the CSE Department of Southeast University, Bangladesh during 2015–2020; Computer Science Department, International Islamic University Malaysia (IIUM), Malaysia, during 2010–2015; at BRACU, Bangladesh, during 2009–2010; and at NSU, Bangladesh, during 2004–2005. He was a guest lecturer for the STEP project in the Department of Technical and Vocational Education, Islamic University of Technology, Bangladesh, in 2018. He also worked as a Researcher at Networking Lab, Kyung Hee University, South Korea, from September 2005 to August 2009 where he completed his MS leading to PhD. His research interests include wireless sensor networks, network security, cloud computing, and e-services technologies. Currently, he is also working on some multidisciplinary issues. He is a recipient of several awards/best paper awards and has several notable publications in these areas. So far, he has delivered over 20 Keynotes and Invited speeches at various international conferences and events. He has served as a General Chair, Organizing Committee Member, and Technical Program Committee (TPC) member in numerous top-ranked international conferences/workshops like INFOCOM, GLOBECOM, ICC, LCN, GreenCom, AINA, WCNC, HPCS, ICA3PP, IWCMC, VTC, HPCC, SGIoT, etc. He was awarded the IEEE Outstanding Leadership Award for his role in IEEE GreenCom'13 conference. He is currently serving as the Editor-in-Chief of *International Journal of Computers and Applications*, Taylor & Francis, UK; Associate Technical Editor of *IEEE Communications Magazine*; Editor of *Ad Hoc and Sensor Wireless Networks*, Old City Publishing, *International Journal of Sensor Networks*, Inderscience Publishers, and *Malaysian Journal of Computer Science*; Associate Editor of *Connection Science*, Taylor & Francis, UK, *International Journal of Computational Science and Engineering*, Inderscience; Area Editor of *International Journal of Communication Networks and Information Security*; Guest Editor of many special issues of top-ranked journals; and Editor/Author of 21 books. One of his books has been included twice in Intel Corporation's Recommended Reading List for Developers, second half of 2013 and first half of 2014; three books were included in IEEE Communications Society's (IEEE ComSoc) Best Readings in Communications and Information Systems Security, 2013; two other books were indexed with all the titles (chapters) in Elsevier's acclaimed abstract and citation database, Scopus, in February 2015; and a seventh book is translated to simplified Chinese language from English version. Also, two of his journal papers and one conference paper were included under different categories in IEEE Communications Society's (IEEE ComSoc) Best Readings Topics on Communications and Information Systems Security, 2013. He also serves as a referee of many prestigious journals. He received some awards for his reviewing activities like: one of the most active reviewers of *IAJIT* several times; Elsevier Outstanding Reviewer for *Computer Networks*, *Ad Hoc Networks*, *FGCS*, and *JNCA* in multiple years. He is a Senior Member of the Institute of Electrical and Electronics Engineers (IEEE), USA.

Deep Learning Approaches for IoT Security in the Big Data Era

K. S. Sunitha Krishnan and Sabu M. Thampi

He who would search for pearls must dive below

John Dryden

1 Introduction

The confluence of innovative technologies in wireless communications led to the evolution of the Internet of Things (IoT). According to recent studies, this cartel of things entrenched with electronic components, software, sensors, actuators coupled with the Internet, will increase to 50 billion by 2020. The giant stride in the number of IoT devices makes them the major genesis of data. IoT is triggering a massive influx of big data. To reap out the maximum efficacy of IoT, the massive amount of data is harnessed and converted to actionable insights utilizing the big data analytics. This makes the Internet of Things more intelligent than mere monitoring devices. Big data and IoT works well conjointly to offer analysis and insights. With the conjunction of the Internet of things, big data analytics shift the computing paradigm to the edges for real-time decision making.

A vast amount of data can be seen in various arenas like oil exploration, health care, social media information, power management etc. The organization and interpretation of these data are very useful in business at all levels. This data is unlayered and unstructured which cannot be used in machine learning algorithms which use

K. S. Sunitha Krishnan
Indian Institute of Information Technology and Management – Kerala (IIITM-K), Kazhakkoottam, Kerala, India

Cochin University of Science and Technology, Kochi, Kerala, India
e-mail: sunithakrishnan.res17@iiitmk.ac.in

S. M. Thampi (✉)
Indian Institute of Information Technology and Management – Kerala (IIITM-K), Trivandrum, Kerala, India
e-mail: sabu.thampi@iiitmk.ac.in

Z. Md. Fadlullah, A.-S. Khan Pathan (eds.), *Combating Security Challenges in the Age of Big Data*, Advanced Sciences and Technologies for Security Applications, https://doi.org/10.1007/978-3-030-35642-2_6

supervised instructions. Deep learning can avoid this drawback since they excel in label-less unsupervised learning especially when it comes to prediction and pattern recognition. Deep Learning algorithms create a layered, and hierarchical architecture of learning and representation of data. They have the ability to recognize the latent features and translate them into useful insights in no time. In this chapter, we are trying to identify, review and analyze the state of art deep learning approaches which contribute to the perpetuation of security in the Internet of Things (IoT). We mainly concentrate on the deep learning techniques aiding in the process of authentication feature extraction and detection of threatening invasions and malware. The chapter concludes with the discussion on the challenges faced while developing the algorithms for IoT networks which are suitable for diverse application scenarios and also provides a glimpse to the future perspectives.

1.1 Big Data and Internet of Things

The concept of big data which is characterized by the three Vs, volume, velocity, and variety is a paradigm which has received wide acceptance in the digital world in the last decade. Having more information, big data gave the opportunity to tackle the problems using completely different approach the use of social networks, online services and the development of open source frameworks expanded the possibility of big data. The advent of Cloud computing which offered scalability even broadened the opportunities of big data. IoT was another promising partner to take the big data to a higher level. The amount of data created and stored took a giant leap in this period with the emergence of the Internet of Things (IoT). The Fig. 1 shows the giant stride in the number of IoT devices used, according to the survey done by the connectivist.com. Techniques in big data analytics have the ability to handle the massive amount of continuous stream of data generated by these devices. Figure 2 picturizes the data flow in the process of insight creation from the data collected by the IoT devices.

1.2 Security in Internet of Things

In spite of the considerable benefits of IoT, it comes along with major security problems which need to be addressed. The global connectivity brings along the innate vulnerabilities and security risks. The vulnerabilities present in the IoT devices acts as the entrance for the adversaries to flare up various attacks in the IoT network [1]. The relationship between the vulnerabilities is exploited by the adversaries to invade into these networks. Therefore the advantages of IoT cannot be reaped out to its maximum without the multiple layers of security which safeguards the interconnected systems and devices. The increased rate of diversities, integrated with the wide scale of IoT systems, amplified the security threats of the current communication

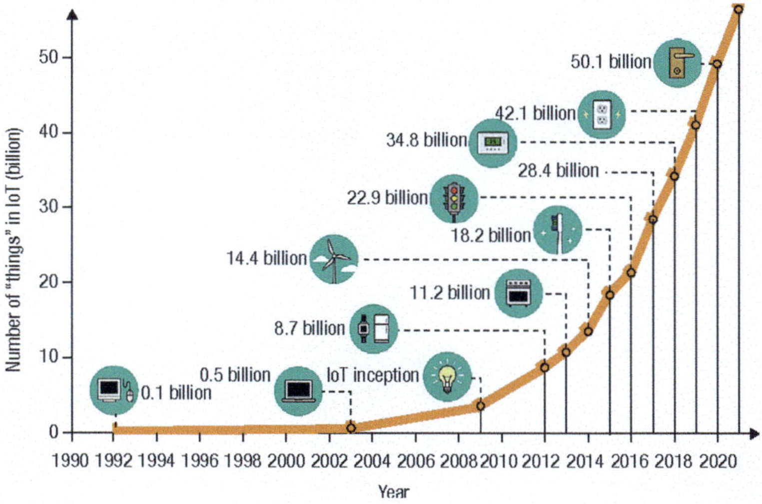

Fig. 1 IoT growth over years

scenario, which is being increasingly used to let interact humans, machines, and robots, in any combination.

IoT infrastructure is engrossed in the continuous transmission of data across the network of things for the achievement of specific goals. In such a symbiotic environment the security requirements authentication, authorization, access control, trust, integrity, confidentiality, privacy, secure middleware and trust etc. needs much importance [2] as shown in Fig. 3. Traditional security measures have to be tailored to the limited resource structure and the ad-hoc nature of the IoT network or new security solutions should be introduced to satisfy these requirements. Additionally, the scalability issue of the structure has to be addressed since the infrastructure is more dynamic in nature. These security issues have to be handled with a high degree of adaptability. Effective security mechanisms are to be deployed befitting to the limited functionality and constraint environment of the IoT systems. This necessitates the techniques for device-level security in communication and network monitoring. Traditional security solutions in conjunction with built-in security in the devices are needed to achieve dynamic detection, prevention, isolation and countermeasures against successful breaches.

Security of Internet of Things spans over all layers of the IoT Infrastructure i.e. Perception layer, network layer, transport layer and application layer [3]. The perception layer consists of various hardware nodes or sensors entrusted with the job of acquisition of various parameters from the residing environment and the network which is responsible for the transmission of the collected data to other nodes. Transportation layer which is an association of heterogeneous networks provides pervasive access atmosphere for the for perception layer, understands the information gathered, handles the transmission of data. The application layer consists of the application support layer as well as the IoT application layer. The support layer is

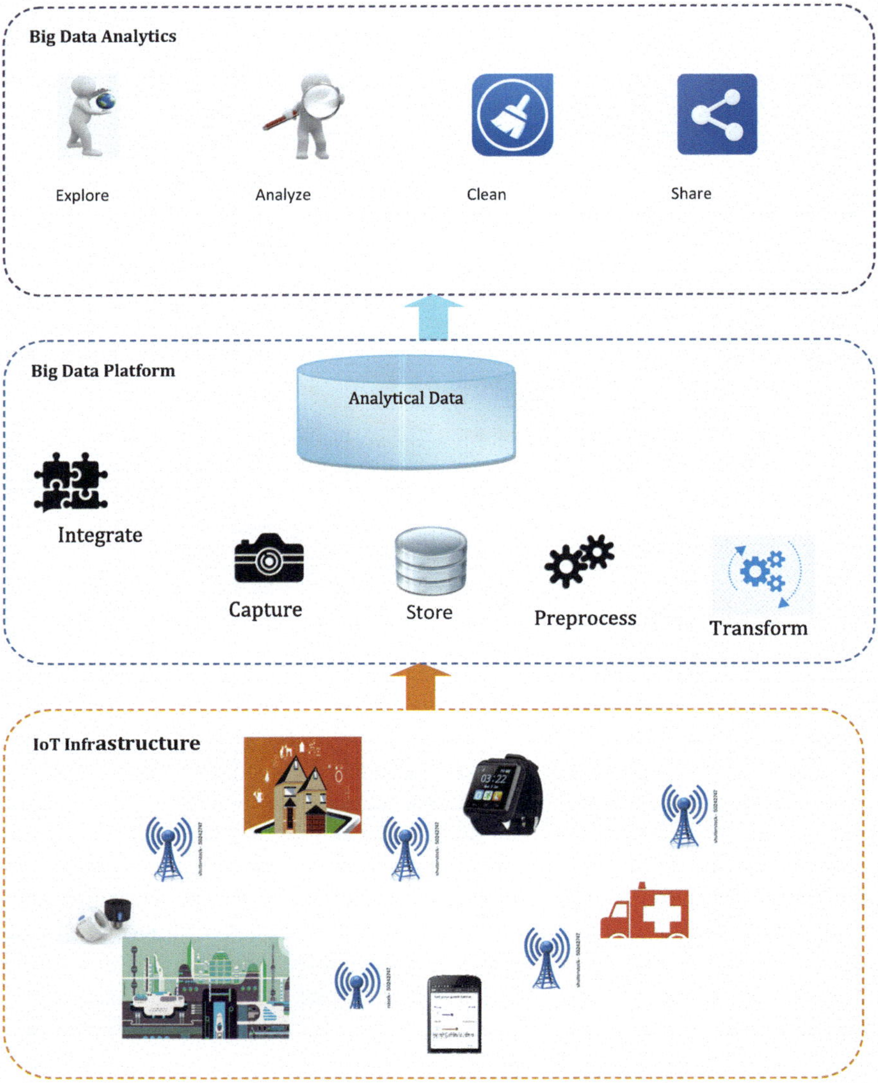

Fig. 2 The flow of Big data in Internet of Things

entrusted with the job of supporting all kind of business services, realizing intelligent computation and the allocation of resources in screening, selecting, producing and processing data. They should recognize malicious and benign data. The IoT application layer includes integrated or specific business applications. The issues in the technologies used in each layer contribute to the security threats of the layer. The summary of the issues in each layer is shown in the Table 1.

The security solutions for these problems cannot be realized by a specific quick fix in a single layer. Therefore, solutions that support cross layer usage are needed

Fig. 3 IoT Security Requirements

Table 1 Summary of security issues in the IoT layers

Sl no	IoT Layer	Security Issues
1	**Perception layer**	No uniform encoding standard for RFID
		Multiple RFID tags send data simultaneously
		Lack of privacy protection
		Lack of trust management systems
		Data confidentiality
		Data authenticity
		Data integrity
		Issues due to heterogeneous integration
2	**Transportation layer**	Access network related issues
		Data security
		Phishing attacks
		Eavesdropping and interference
		Illegal node access
		DDos/dos attacks
		User information leakage
3	**Application layer**	Service interruption and attack issue
		Insecure data
		Issues related to access control
		DDoS attack

to be designed for addressing the issues in IoT infrastructure. The resource constraints in the nodes of a sensor network and multihop communications in open wireless channel make the security of sensor networks even more heavy challenge. Due to the explosive demand of IoT devices and their applications nowadays, the aspect of security demands high priority.

1.3 Overview of Deep Learning Techniques

Deep Learning has been a major focus in data science due to its capability to handle the enormous amount of data tactfully. The key benefit of deep learning in big data is that they can learn from a massive amount of unsupervised data or raw data which is uncategorized. Deep learning in its initial phase was proved to be successful in feature learning tasks. Deep Learning acquires the features itself, which enables the learning process to be more accurate and helps in the creation of better models. Feature extraction using deep learning techniques annex nonlinearity to the data analysis and make the discriminative tasks closely to heuristics. They fit perfect in the IoT paradigm which involves a large amount of data and complex relationships between different parameters, for solving intuitive problems. Nowadays, the potential for deep learning is utilized for classification tasks like intrusion detection, malware analysis, authentication etc.

Deep learning has earned success since it needs very little engineering by hand utilizing large amount of data. According to the authors of [4] a deep-learning architecture is a "multilayer stack of simple modules, all (or most) of which are subject to learning, and many of which compute non-linear input–output mappings". Each node in the stack converts the input to increase both the selectivity and the invariance of the representation. With multiple non-linear layers a system can implement extremely intricate functions of its inputs that are simultaneously sensitive to minute and insensitive to large irrelevant variations.

1.3.1 Evolution of Deep Learning

Deep learning finds its roots in neural networks which were formulated by Walter Pitts and Warren McCulloch in 1943. This mathematical model mimicked the working of neurons, the cells in the human brain which helps them in the thought process and decision making. The 50s and 60s saw the development of machine learning programs and the groundwork of deep learning was put in by Frank Rosenblatt in 1957 with the idea of perceptrons. In 1960 the control theory was introduced by Henry J Kelly, which laid the basics for the development of backpropagation model. The creation of Neocognitron, an ANN mainly used for pattern recognition tasks was created by Kunihiko Fukushima. This model which was used for handwritten character and pattern recognition tasks, recommender systems etc. influenced Hubel and Wiesel which resulted in the formulation of a variant of multilayer perceptrons which needs a minimal amount of preprocessing, Convolutional Neural Networks in 1979. Subsequently, Recurrent Neural Networks which work well for sequential data was introduced in 1980 but gained popularity after the advent of GPUs because of its computational complexities. Later, with the significant progress of backpropagation in the 70s, Yann LeCun combined Convolutional Neural Networks with back propagation in 1989. Long short-term memory a framework of recurrent neural networks was developed in 1997, by Sepp Hochreiter and Juergen Schmidhuber, which works well for sequential data.

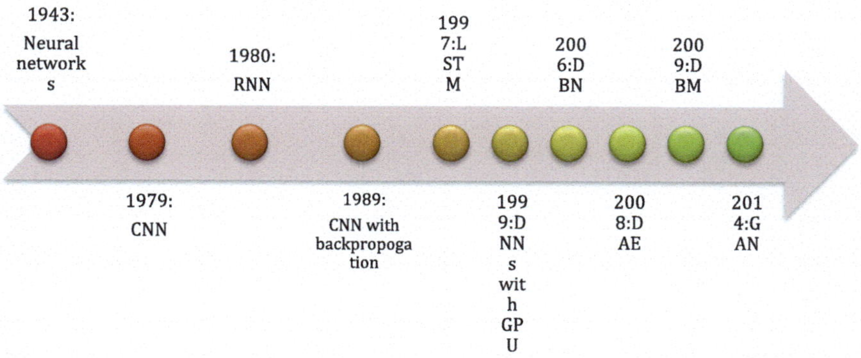

Fig. 4 Evolution of deep learning

With the rise of fast computing GPUs in the late nineties, deep learning took a new dimension. With GPUs faster processing with the images, the computational speed increased by a thousand times. Deep Belief Networks, which is widely used for dimensionality reduction for unsupervised training and as a classifier for supervised training, were introduced by Hinton in 2006. 2008 saw the emergence of Denoising Auto encoders which is trained to build up the data from the input containing noise. Deep Boltzmann Machine, in which the output of one BM is cascaded to multiple BMs, was introduced in 2009 by the Hinton. Recent advancement in deep learning is the introduction of Generative Adversarial Networks (GAN) which comprises two networks competing for each other to learn the data and get smarter. It is considered the most interesting idea in the last ten years of machine learning. Figure 4 shows the time line of the various mile stones in the development of deep learning techniques. Deep learning acts as a central axis where the processing of Big Data and the evolution of Artificial Intelligence, revolve around. Deep Learning is still in its adolescence and needs many innovative ideas to be incorporated.

1.3.2 Deep Learning Architectures

Deep learning has become one of the hot topics of research in the area of artificial intelligence. We present various deep learning architectures and their brief descriptions. Figure 5 shows the broad classification of deep learning algorithms.

Unsupervised(Generative) Algorithms

Unsupervised (generative) algorithms make the most of unlabelled data for training. They learn the likelihood of a given input to be in a class label and are assigned to the label to which it has the highest probability. Following sections explain various types of unsupervised mechanisms.

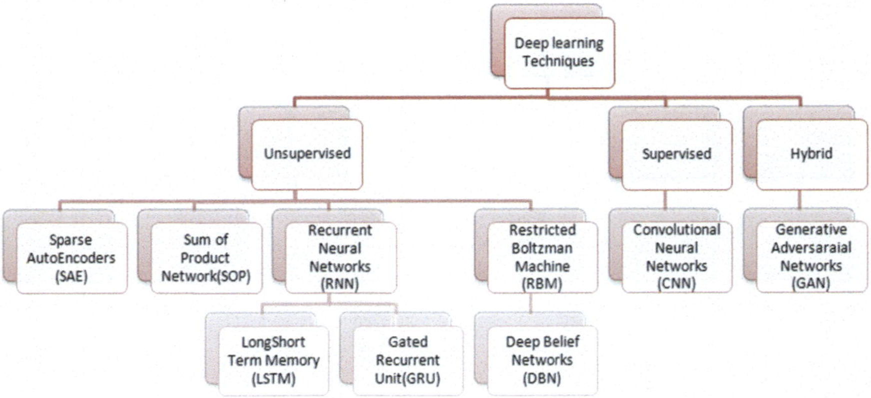

Fig. 5 Classification of Deep Learning algorithms

Sparse Auto Encoders

Auto encoders are neural networks which apply back propagation and try an approximation to the identity function such that the output x^ $\approx x$, where x is the input. The identity function is a trivial function applying several constraints like limiting the number of hidden units on the network, discovering inherent structure about the data. Auto encoders have an encoding stage and a decoding stage [5]. In the encoding stage, the input x is converted to the hidden layer h using the encoding function h.

$$h = f\left(W(1)x + b(1)\right).$$

Then the hidden representation h is recreated to the original input in the decoding stage.

$$y = g\left(W(2)h + b(2)\right).$$

Stacked (Sparse) auto encoders can be considered as a deep learning model which is constructed by stacking multiple auto encoders (as shown in Fig. 6) which uses layer-wise unsupervised pre-training. Pre-training in Auto Encoders is to train a single auto encoder using a single hidden layer. Each Auto encoder is trained separately before cascading it [6]. The number of nodes in hidden layers of the auto encoders will be lesser than that in the input layer which represents a new reduced feature set. The data is then reconstructed after complicated computations and these new transformed features are formed at different depths in the network. Denoising Auto encoders are a variant of auto encoders that is trained to build up the data from the input containing noise.

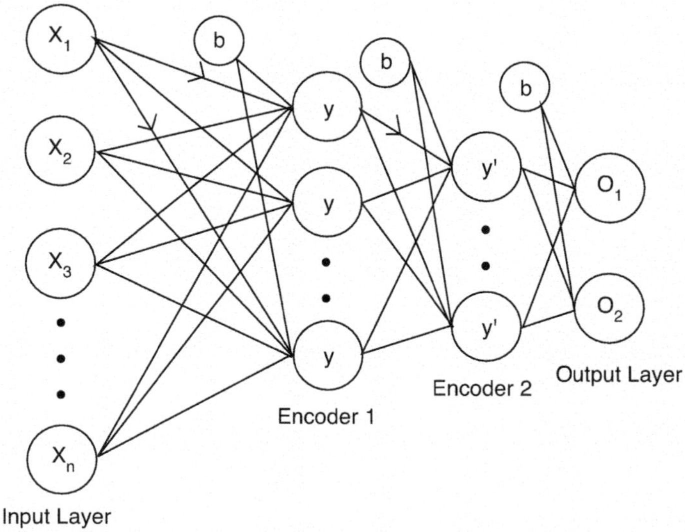

Fig. 6 Sparse Auto Encoder with two hidden layers and two class labels

Fig. 7 SPN implementing a naïve Bayes mixture model

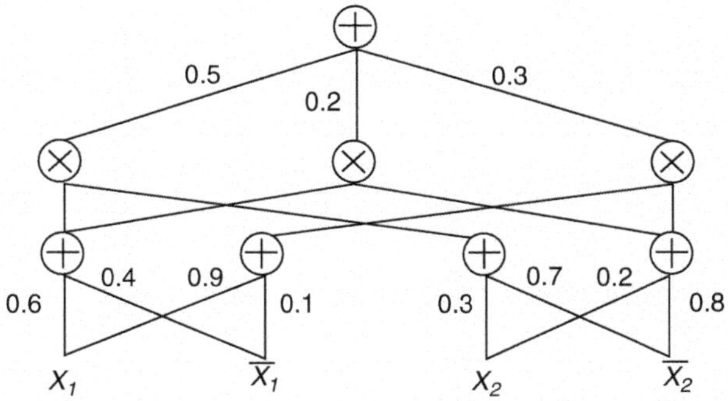

Sum of Products Network

Sum of Product network(SPN) is a deep probabilistic model representing a tractable probability distribution [7]. They can incorporate features into an expressive model without requiring approximate inference. It is a rooted directed acyclic graph whose leaves are the variables and whose internal nodes are sums and products [8]. The sum nodes provide mixture models, while the product nodes express the feature hierarchy. Figure 7 shows an example of an SPN implementing a naive Bayes mixture model with three components and two variables. SPNs have achieved remarkable results on numerous datasets.

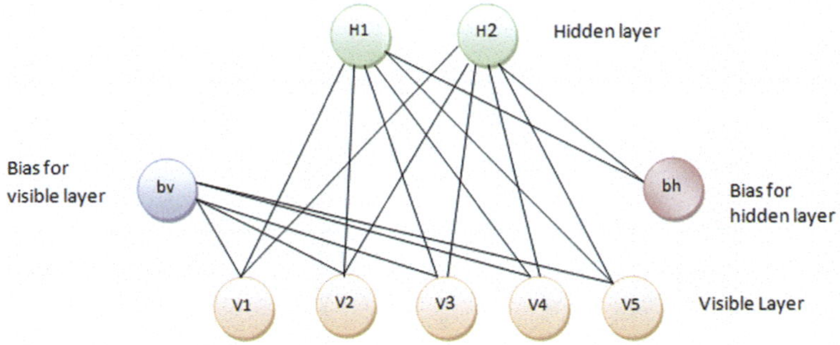

Fig. 8 RBM architecture

Restricted Boltzmann Machine

Restricted Boltzmann Machine (RBM), a network of stochastic neurons is a part of the family of energy-based models. At the same time, it is a probabilistic model too. They have the easiest architecture with two layers, visible layer and the hidden layer, and bias for each layer. Figure 8 shows the schematic representation of RBM. The hidden layer takes part in the process of transformation in the system maintaining the impervious to the observations. The neurons in the machine are in binary state 0 or 1 in a particular point of time. The state refers to the values of neurons in the visible and hidden layers. Conditional Probability is calculated for each node at each state P(h|v) to calculate the value of each unit in the hidden layer and then uses the conditional probability P(v|h) to calculate the value of each unit in the visible layer. This is repeated until convergence.

Deep Belief Networks

Deep Belief networks were constructed by Hinton by stacking various Restricted Boltzmann Machines creating a generative model consisting of many layers by greedily training each layer (from lowest to highest) as an RBM using the previous layer's activations as inputs (Fig. 9). The RBM in each layer exchanges the information with both the former and subsequent layers. Each layer is made up of a set of binary or real valued units. The heap of RBMs has a final Softmax layer which makes it a classifier that groups the unlabeled data in an unsupervised manner. Other than the initial and the final layer in Deep belief networks every layer serves as hidden layers to the nodes comes prior to them and as input (visible) to the nodes that come later [9].

Recurrent Neural Network

Recurrent Neural Networks are conventional sequential learning models that are effective in the processing of sequential information. They are called recurrent networks since they carry out the same job for every input independent of the prior

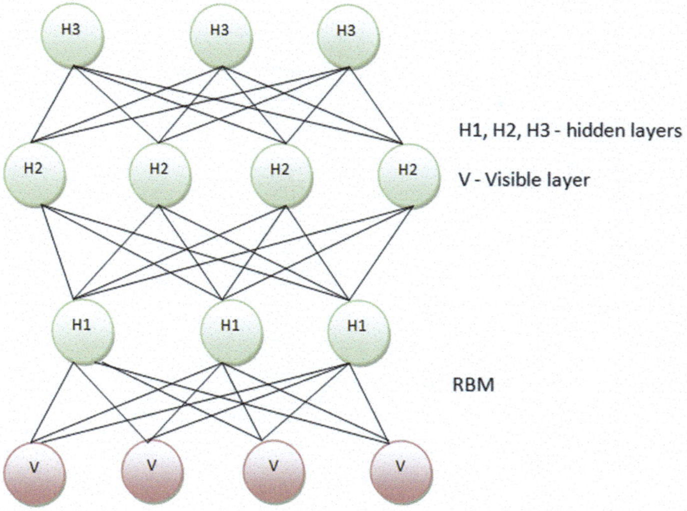

Fig. 9 DBN architecture

computations. They learn the features for the data by keeping the former inputs in the memory. A directed cycle is brought in to create the connections between neurons, as shown in Fig. 10 [10]. The deepness of the network will be as large as the length of the input data sequence. RNN has been found more beneficial in modeling the sequential data.

The input units: $\{x_0, x_1, \ldots, x_t, x_{t+1}, \ldots\}$
The output units: $\{y_0, y_1, \ldots, y_t, y_{t+1}, \ldots\}$
The hidden units: $\{H_0, H_1, \ldots, H_t, H_{t+1}, \ldots\}$.

At the time step t, the recurrent neural network takes the current sample x_t and the previous hidden representation H_{t-1} as input to obtain the current hidden representation

$Ht = f(x_t, H_{t-1})$, *f* is the encoder function

Several RNNs can be piled together to get a deep learning model. RNNs and its variants have displayed impressive performance in the domains like speech recognition, natural language processing etc. where there exists dependency among the input data.

Long Short term Memory

Recurrent neural networks capture random length dependencies of the input data but fail to acquire long-term dependencies because of the vanishing gradient. This drawback is surpassed by the model long short term memory model introduced by Hochreiter and Schmidhuber by preserving the error forbidding the gradient explosion. LSTM is a variant of RNN with four neural networks in a single layer. The main feature of LSTM is the presence of the state cell on the top of every layer,

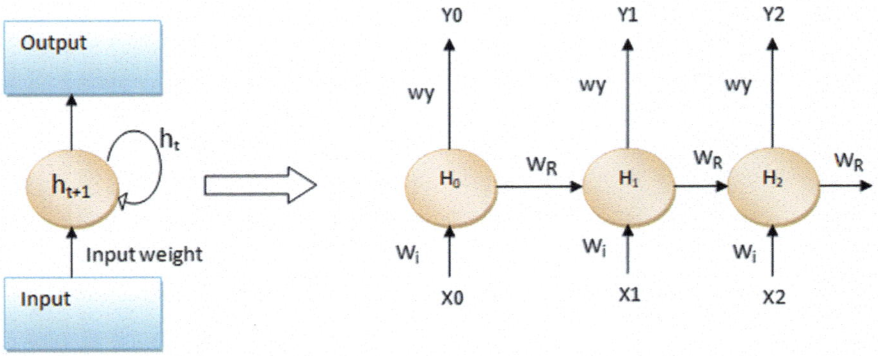

Fig. 10 RNN Architecture: Unfolded (right)

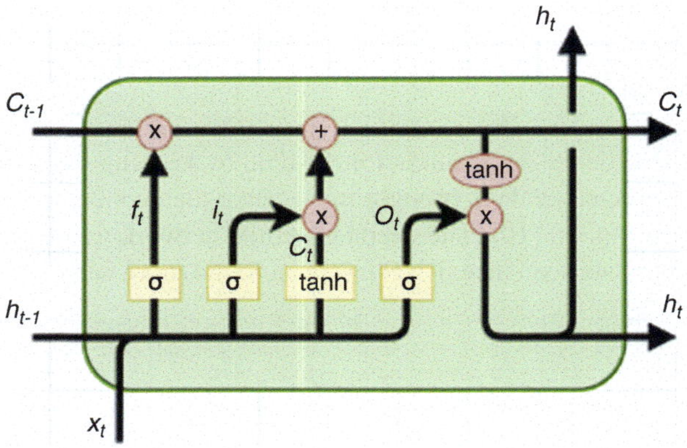

Fig. 11 Architecture of LSTM

which is responsible for the transmission of information from the former layer to the next layer. The gates in the LSTM accounts for the management of the information to be passed or dropped. To control the flow the information input gate, forget gate and output gates are used as shown in Fig. 11.

Gated Recurrent Units

Gated recurrent Unit is a less complex model of LSTM model decreasing the number of gates in the architecture. The GRU combines the "forget gate" and "input gate" in an LSTM to form an "update gate" and merges the hidden state and cell state, which led to the formation of a much simpler architecture of the model as shown in Fig. 12.

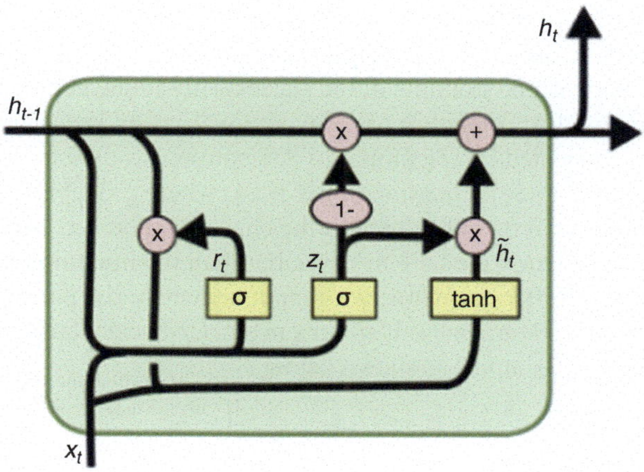

Fig. 12 Architecture of Gated Recurrent Unit

Supervised Learning

The main aim of supervised learning or discriminative model is to distinguish some parts of data for pattern classification with labeled data. Convolutional Neural Networks is the discriminative model among the deep learning models.

Convolutional Neural Networks

Convolutional neural networks are the deep learning model used extensively for feature learning and image classification. The main reason for the drastic boom in deep learning was the use of Convolutional networks in image recognition. They are used to categorize the images, group them by the similarity and do object recognition within the images. These algorithms had the ability to identify faces, persons, street signs and many other variations of perceptible data. Analogous to the other traditional neural networks, its structure is influenced by the neurons in animal and human brains. It mimics the visual cortex in a cat's brain containing a complex sequence of cells.

The time delay networks were the key influence on the origin of CNN. The reduction in the computation in TDNNS is due to the fact that the weights are been shared in a temporal dimension. The matrix multiplication in the conventional neural networks was replaced by convolutions in Convolutional neural networks. Thus the complexity of the network was reduced with the reduction of a number of weights. The feature extraction process in the traditional learning algorithms refrains in these networks thereby the images can be directly fed into the networks as raw input. So minimal preprocessing is done in the case of CNN model. Spatial relationships are utilized to reduce the number of parameters in the network, and leveraging the standard back propagation algorithms the performance is improved. Multilayer networks can be trained by CNN utilizing gradient descent to learn complex, high dimensional non linear mappings from large collections of data. Three basic concepts, local receptive fields, shared weights, and pooling is used by these networks. AlphaGo by Google is one example of the successfully implemented using CNN.

CNN is composed of a number of Convolutional layers succeeded by pooling layers and fully connected layers(similar to Perceptrons) as final layers as shown in Fig. 13. The input is three dimensional, p x p x q where p denotes the height and width of the input, q refers to the depth of the channel. There exist several filters in each layer of size m x m x n where m is smaller than the input image but n can be lesser or equal to q. Filters convolve with input and share the parameters, weight, and bias to create the feature maps of size p x m x 1. CNN calculates the dot product with the weights and its inputs as shown below

$$h^k = f\left(W^k\, X\, x + b^k\right)$$

But the created inputs are small regions of the real input volume. Overfitting is controlled by decreasing the parameters in the network by down sampling the feature map. A small contiguous region of the filter size is selected and the pooling operation is done on the region. Pooling might be max pooling or average pooling. Similar to the traditional neural networks the final stage layers are fully connected layers. They produce a high level abstraction of the data utilizing the prior low level and mid-level features. The final layer produces the probability of an instance to in a specific class or the classification scores.

For the classification of images as in the case of Fig. 13, the raw pixels will be the input to the Convolutional neural network. CNN learns to discover the edges from these raw pixels in the first layer. It utilizes these edges to identify simple shapes in the next layer. The successive layers will be capable of learning higher level features like facial shapes, buildings etc. utilizing the simple shapes from the previous layer.

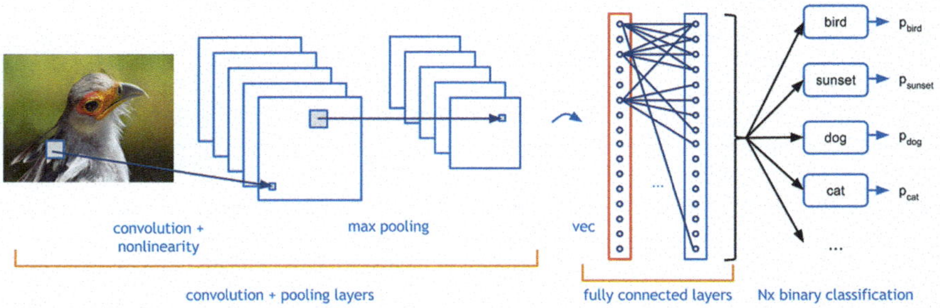

Fig. 13 Architecture of CNN

Hybrid Learning

Hybrid architecture integrates the advantages of supervised and unsupervised learning. They try to cluster data as well as identify the data. Generative adversarial network is an example of hybrid learning technique.

Generative Adversarial Networks

An innovative framework which trains both supervised and unsupervised simultaneously was put forward by Goodfellow in 2014 [11]. It consists of a two models generative G and discriminative D as shown in Fig. 14 where G captures the distribution of the data p_g in the real data *t* and D model differentiate the original input data and the data from the model G i.e. p_m. In every iteration, the generative model is opposed against its adversary, a discriminative model which tries to identify whether the given sample is generated by the model or the original data. Generative Model G generates more realistic data to fool and complicate discriminator model D tries to identify the genuine ones. Tug of war among these models helps them improve their techniques to identify the genuine one from the fake one. This two-player game is conclusively proved with Value function V(G,D).

$$\min_G \max_D V(G,D) = E_{t-pdata}[logd(t) + E_{m-pm(m)}\left[\log\left(1 - D\left(G\left(m\right)\right)\right)\right]$$

Where

D(t): the probability that t came from the data
p_{data}: distribution of the real-world data.

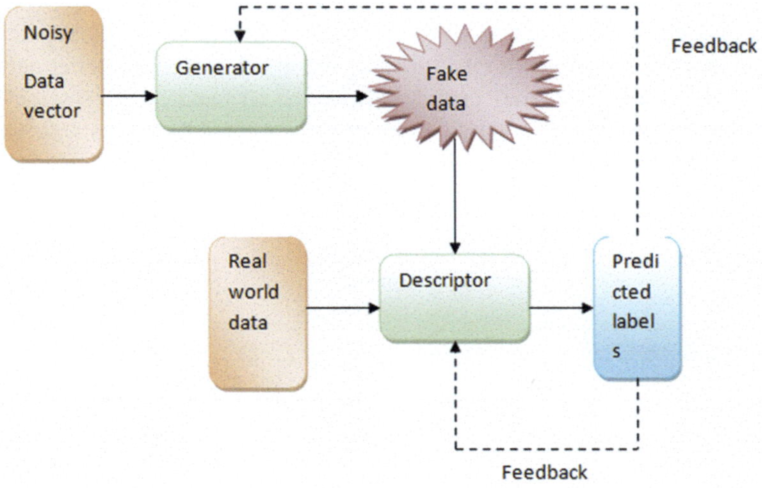

Fig. 14 Architecture of GAN

The model reaches the equilibrium when both reach the point where none of them can be improved i.e.

$p_g = p_{data}$. This means that the discriminator can no longer identify between the two distributions.

2 Pertinence of deep learning in IoT security

Deep learning techniques which gained remarkable achievements in the area of computer vision, automatic speech recognition, pattern recognition etc., have now been used extensively for the sustainment of security in IoT [12]. Figure 15 shows the taxonomy of the application of deep learning techniques for IoT security. They are classified as the approaches used for authentication, intrusion detection, feature selection and malware detection.

2.1 Deep Learning for Authentication

The focus of the deep learning techniques while applied in authentication is on identity assurance rather than fraud detection. These techniques have been used for the authentication of the users as well as the IoT devices. Various deep learning approaches used for the authentication process of users and IoT devices are summarized in Table 2.

2.1.1 User Authentication

A user authentication framework was proposed by Lee et al., extracting features based on users' interaction with the touchscreen, which used deep belief networks to classify the users [13]. They extracted stroke based features and session-based features for authentication. A modified DBN with two hidden layers was used for

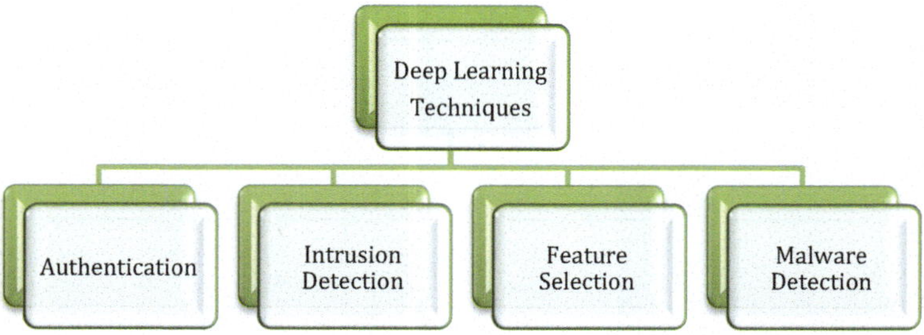

Fig. 15 Applications of deep learning in IoT security

Table 2 Deep Learning Techniques for Authentication

Sl No	Deep Learning Techniques for Authentication	Inferences
1	User authentication using stacked auto encoders [15]	Authentication of users from the physiological and behavioral features from channel state information (CSI) measurements of WiFi signals.
		Resilient to user spoofing attacks.
2	User authentication using Deep Belief Networks [13]	Collects user interaction features, stroke based features and session based features to authenticate the user.
		DBNs including the dropouts are used thereby avoiding overfitting on small training sets.
3	User authentication and identification using Deep Belief Networks [14]	Provide very less EER in identifying and authenticating a user using keystroke timing features.
4	Authentication of IoT devices using LSTM [17, 18]	Preserve long term dependencies in sequential data which are suited for wireless signals.
5.	Signal Authentication using LSTM [17]	LSTM reduces complexity and latency of the attack detection compared to other security methods
		Authenticates the signal, extracting the stochastic features from IoT signal and watermarking these features inside the original signal.
		Allows the cloud to detect sophisticated eavesdropping
		Attacks, since the attacker will not be able to extract the watermarked information

the classification. DBN produced impressive results compared to other examined methods with an identification rate of 81.5% and a median EER of 9.93%. Deep Belief Network was used for authentication using another modality, keystroke dynamics (a behavior-based unique timing patterns in an individual's typing rhythm) by Saket et al. [14]. They considered the identification of a user as a binary classification problem and used the keystroke features like the hold time, key down- key downtime and key up-key downtime to authenticate a person. This network model, DeepSecure had three hidden layers of 100, 400 and 100 dimensions layers. The considerable number of hidden layers introduces sparsity which can secure the inter-feature relations. This eliminates the need for manual feature engineering and eventually bringing forth a model which is more robust and less prone to over fitting on this key-stroke recognition problem when compared to a simpler 1 hidden layer model. Another deep learning based user authentication Scheme (as shown in Fig. 16) was proposed by the authors of [15], in which representative features were extracted from channel state information (CSI) measurements of Wi-Fi signals, to accurately identify an individual user. The system performs activity recognition and human authentication by building a three-layer deep neural network (DNN) model based on AutoEncoder. Unlike other authentication schemes based on high dimension feature sets and linear classification models (e.g., SVM), non-linear physical and biometric abstractions learned by DNN model are computation efficient and are robust to small-scale input variations. The stated network roughly identifies the

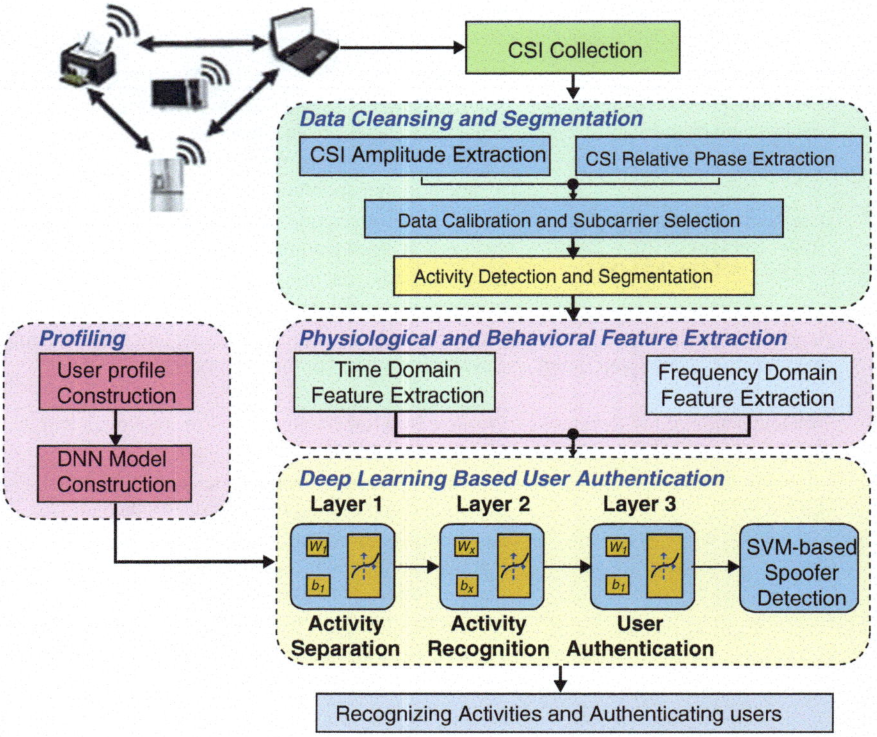

Fig. 16 Overview of deep learning based user authentication formulated by Shi et al., 2017

activity type in the first layer and subsequently the activity details in the second layer. The third layer recognizes each individual user with a softmax function. The integration of the SVM model with the DNN, immune the system against spoofing attack.

2.1.2 Device Authentication

The efficacy of the model LSTM was exploited by Rajshekar et al. to learn the hardware imperfections of the low powered radio device and identify the features which makes them unique [16]. LSTM capitalize on the temporal correlation between the I/Q streams of wireless signals to identify legitimate nodes from high power adversaries that transmit identical modulation, coding, and even data, given that these adversaries inadvertently introduce their own distinct imperfections. The technique was examined with LoRa transmitters and much higher software radio Adversaries and found buoyant to noise, multi-path, and signal attenuation. This approach was strengthened by Ferdowsi et al. by integrating game theory to the framework [17].

The framework was designed to allow the cloud to authenticate the signals and disclose the presence of any adversary who may change the devices' output signal. LSTM extracts the stochastic features like spectral flatness, skewness, kurtosis, and central moments from IoT signal and watermark these features inside the original signal. Since enormous amount of computational resources is required for the authentication, the cloud cannot authenticate all transmitted signals from the IoT devices simultaneously. Predicting the vulnerability of IoT devices is considered as a non-cooperative game between the cloud and the attacker, considering the constraint of the resources in these devices. The cloud optimally chooses the device to be authenticated with the help of Nash equilibrium. 30% reduction in the number of compromised devices was observed using this approach and improved the protection of the system in massive IoT scenario.

2.2 *Deep Learning for Intrusion Detection*

On top of the secure foundation built by the cryptographic techniques and the secure protocols Intrusion Detection Systems (IDS) act as the first layer of defense in the arena of IoT Security. The ability to recognize, the patterns of typical attacks and abnormal activity patterns, makes IDS primary choice which can be deployed over all levels. IDS monitors, recognize the patterns of typical attacks and abnormal activity patterns and reports to the security management system. Deep learning which has been considered as a breakthrough in the arena of Artificial Intelligence has raised the potential of intrusion detection to achieve high detection rate and low false alarm rate. They utilize the network traffic data to identify the intrusions. Tables 3 and 4 summarizes the deep learning approaches used for the classification tasks in intrusion detection systems.

Most of the literature has utilized the KDDCUP99, NSL–KDD and UNSW-NB15 datasets to substantiate their proposed techniques. KDDCUP99 is a collection of raw TCP dump data which contains 41 attributes and a label assigned to each instance as either attack type or as normal. There are 22 attacks in the training data out of the 39 attacks in the test data. The attack types are categorized into 4 groups: DOS: Denial of service – e.g. syn flooding, Probing- Surveillance and other probing, e.g. port scanning, U2R: unauthorized access to local superuser (root) privileges, e.g. buffer overflow attacks, R2L: unauthorized access from a remote machine, e.g. password guessing. NSL-KDD is a sophisticated version of KDDCUP99 having similar attack types. UNSW-NB15 data set is also raw traffic dataset which contains 9 attack groups- Backdoor, Analysis, DoS, Exploits, Fuzzers, Generic, Reconnaissance, Shellcode and Worms. There are 49 features in the dataset along with the class label. The deep learning architectures act on these raw traffic data to categorize benign traffic and attacks.

Table 3 Deep Learning approaches for Intrusion detection

Sl No	Approaches for Intrusion detection	Inferences
1	Deep neural networks for Intrusion detection [21] [19]	Effective attack detection
2	Distributed attack detection scheme using DNN([20]	Better performance than the centralized model
		Collaborative sharing of learning parameters avoids overfitting of local parameters which helps in achieving better performance
3	Deep neural networks for intrusion detection in invehicle security [23]	A real-time response to the attack with a improved detection ratio in controller area network (CAN) bus.
4	Ensemble Algorithm using DNN for Intrusion Detection [24]	DNN along with spectral clustering algorithms is used.
		Better performance than shallow counterparts
5	Intrusion detection using Auto Encoders [25]	Self taught learning
		Works on unlabelled network traffic data collected.
6	Stacked auto encoders for Intrusion detection [6].	IDS designed for different layers uses layer specific features.
		Lightweight IDS achieve comparable detection rate as the ordinary IDS.
7	Auto Encoders for traffic Identification [26]	Deep structures of works better than shallow counterparts.
8	Intrusion detection using Deep Belief Network [27]	Improved classification rate for known and unknown attacks with minimum number of false alarm rate.
		Achieved higher accuracy with the training done on smaller amount training data
9	Deep belief Network for Intrusion detection [28]	Uses four hidden layer RBMs.
		Efficient use of very large sets of unlabeled data and can be pre-trained in completely unsupervised learning.
		Limited labeled data is used to fine-tune DBN for a classification task
10	Restricted Boltzmann machine for Intrusion detection [29]	Combines the expressive power of unsupervised models with good classification accuracy capabilities to infer part of its knowledge from incomplete training data.
		Not restricted to a prior knowledge base,
		It can enable the detection of any type of unknown anomalous events
		Effective in coping with the zero-day attacks.

2.2.1 Deep Neural Networks

Far-reaching researches have been done in the area of detection of intrusions in the cyberspace. A deep learning method was used by Yavuz et al. to identify the routing attacks in the Internet of Things [19]. They used highly scalable, deep-learning based attack detection methodology for detection of IoT routing attacks with high accuracy and precision for continuous monitoring. Another attempt using the same

Table 4 Recurrent Neural Networks for Intrusion detection

Sl No	Approaches for Intrusion detection	Inferences
1	LSTM-RNN for intrusion detection [35]	Find an optimal hyper-parameter for LSTM-RNN and confirm the detection rate and false alarm rate
2	Reduced size recurrent neural networks for Intrusion detection [32]	a reduced-size structure of RNN is used, based on the four group of input features
		Improved classification rates
3	Recurrent neural networks for Intrusion detection [33]	Fully connected model has stronger modeling ability and higher detection rate than the reduced-size RNN model
4	Deep recurrent neural Network paradigm for intrusion detection [10]	Uses a descriptive model of deep recurrent neural Network (RNNs)
		Works with low false alarm with new unseen threats
		Bi-directional techniques can neutralize sequence dependencies via considering forward and backward order of request sequences.
5	LSTM-RNN for DDoS attack detection [36]	Resolve the vanishing gradient problems
		Keep details of attacks learnt from training process and make detection decisions based on this stored information on gated cell
6	LSTM based ensemble method for intrusion detection [37]	A language based model for intrusion detection
		Learns the semantic meaning and interactions of each system call
		Needs significant smaller training overhead since no database is used for the storage of patterns
7	MS-LSTM for anomaly detection [40]	A multiscale LSTM is used assuming the internet flow as a multi-dimensional time sequence and learns the traffic pattern from historical features in a sliding time window.
8	Traffic classifier using CNN-RNN [41]	RNN combined with a convolutional neural network is used to provide best detection results.
		Robust and gives excellent F1 detection scores under a highly unbalanced dataset
9	Gated recurrent unit for Intrusion detection [39]	Bi-directional GRU and multi-layer GRU is used for intrusion detection
10	CNN –RNN for Intrusion detection [42]	CNN with first layer and variants of RNN are used as subsequent layers.
		Remarkable performance than other classifiers.

approach was done by Diro et al. [20] for identifying the intrusions in the IoT network. They utilized the self taught and data compression capabilities of deep learning techniques to discern attacks from the benign traffic. They proved that the deep model outperformed the existing methods available for detecting the attacks. Impressive detection rate was observed for the experiments with the stated approach [21]. An ensemble model which combines spectral clustering with deep neural networks to detect the attack types was proposed by Ma et al. [22]. The Clusters capture the network features and break down them into k subsets to learn more knowledge and patterns from analogous clusters. Deep neural networks help in acquiring highly abstract features from these subsets. The model is proficient in classifying the sparse attack cases and increases the security in real security systems. The optimization of the weight parameters and the thresholds of each DNN layer remain as a limitation of the work. DNN has been applied to secure the in vehicular network inspecting the CAN network packets. Experimental results demonstrate that the stated method demonstrates a superior performance in terms of the detection rate.

2.2.2 Auto Encoders

Auto encoders are widely used for dimensionality reduction and data denoising nowadays. But attempts were done for the classification tasks too. Niyaz et al. used self-taught learning, based on sparse auto-encoder and soft-max regression, to develop a Network Intrusion Detection Systems [25]. Auto encoders were used to learn the features from the dataset. The learned features were applied to the labeled test dataset for classification. They used the n-fold cross-validation technique for the evaluation of performance and obtained a reasonable result. Aminanto et al and Wang et al. have examined the applicability of autoencoders as classifiers in network traffic data [6, 26].

2.2.3 Restricted Boltzmann Machines

The capability of Restricted Boltzmann machines to identify the latent factors in the data is exploited to find the anomalies in the security domain. The abnormal behavior of the network depends upon several factors and these factors are captured easily by Boltzmann machines and classify them as benign traffic or attack traffic. Fiore et al. utilized RBM which belongs to the family of energy based models to find anomalies in the network in a semi-supervised manner [30]. The generative power and the classification accuracy of DRBM make them efficient to extract the inherent aspects of the benign traffic. Since they are not confined to any prior knowledge base, they can be used for the detection of anomalous behavior. The performance of such IDS is enhanced by combing RBM with SVM by Bo Dong et al. [31].

2.2.4 Deep Belief Networks

Being the most influential deep neural networks, DBN is used for classification while associating the class labels with the feature vectors [27]. DBN utilize a very large set of unlabeled data and make use of unsupervised learning for pretraining. A limited number of labeled data can be used for the process of fine-tuning the model for classification. Gao et al. have proved that the deep belief networks perform better than the SVM and traditional neural networks [28].

2.2.5 RNN

RNN are powerful for modeling sequences since they have cyclic connections. Sheikhan et al. proposed a reduced-size structure of RNN, based on the four group of input features. They showed remarkable classification rates [32]. However, the nodes of layers are partially connected, the reduced RNNs do not show the ability of deep learning to model high-dimensional features, and the authors do not study the performance of the model in the binary classification. Chuan et al. proposed a three layer RNN architecture with 41 features to model a deep approach for intrusion detection [33]. They proved that fully connected model has stronger modeling ability and higher detection rate than the reduced-size RNN model and is superior to other classification methods in both binary and multiclass classification.Lopez et al. used a combination of RNN with CNN to classify the network traffic [34].

Jihyum etal proposed LSTM- a variant of RNN based model for finding the intrusions [35]. They find an optimal hyper-parameter for LSTM-RNN and confirm the detection rate and false alarm rate. Two variations of RNN, bi-directional RNN, Long Short Term Memory (LSTM) and bi-directional (LSTM) was used by Elsherif to develop solution that detects anomaly inside a sequence of user's requests [10]. He used a descriptive model of deep Recurrent Neural Network (RNNs) and works with low false alarm with new unseen threats. He proved that bi-directional techniques can neutralize sequence dependencies via considering forward and backward order of request sequences. The problem of vanishing gradient is resolved by using LSTM for IDS by Bediako et al. [36]. It keeps the details of attacks learnt from training process and make detection decisions based on this stored. A language based ensemble model for intrusion detection was proposed by Kim et al. which learns the semantic meaning and interactions of each system call [37]. It needed significant smaller training overhead since no database is used for the storage of patterns. Cheng et al. used a multiscale LSTM assuming the Internet flow as a multi-dimensional time sequence and learns the traffic pattern from historical features in a sliding time window [38]. Gated Recurrent Unit has been used to detect the attacks by the authors of [39]. They used bi-directional GRU and multi-layer GRU for intrusion detection.

2.2.6 Convolutional Neural Networks

Convolutional Neural Networks have been widely used in the field of computer vision since they have proved its efficacy in working with the images. A small amount of work in the area of intrusion detection is available in the cyber security paradigm using CNN. The capability of CNN to excerpt high-level feature representations that portrays the abstract form of low-level feature sets of network traffic is exploited to distinguish benign and malignant traffic. The authors of [42] assessed the efficacy of CNN and the integration of sequential data modeling techniques for the classification of benign and malignant network connections. They used CNN as the first layer with a recurrent neural network and its variant as subsequent layers.

They claim that deep learning based approaches such as CNN and RNN, LSTM, GRU are suitable at modeling network traffic as a sequence of TCP/IP packets in comparison to other conventional machine learning classifiers. Lopez et al. used a combination of RNN with CNN to classify the network traffic [34].

2.3 Deep Learning for Feature Selection

Feature Selection is a major process that influences the performance of a specific model. The absence of manual feature manipulation is one of the important advantages of deep learning. Deep learning has been used for feature selection for an intrusion detection system by the authors of [6] to prove that the reduced input features are sufficient to achieve comparable detection rate as the whole features. They provide good feature representation of the unlabelled data collected from the network. Wang et al. have used stacked auto encoders to learn an efficient, compressed representation for a set of data for the identification of anomalies in TCP flow data [26]. The three-layered architecture transformed the raw data very efficiently with some computations in an unsupervised manner which makes them the prime choice for feature extraction. Auto encoders can restore data based on less information loss and error. Li et al. used this approach to prepare the data for malicious Code Detection, converting high-dimensional data into low dimensional codes with the nonlinear mapping and extracted the main features of the data [24]. Tobiyama et al. utilized the efficacy of RNNs with LSTM units to construct behavioral language model for the extraction of features from the process behavior of the terminal [43]. The model consists of an input layer a normal hidden layer, two hidden LSTM layers, and an output layer. Dropout for non-recurrent connection is used in the training phase. The features processed using trained RNN are then converted to feature images. The information of previous inputs is accumulated in the last hidden layer. Some sort of regularity will be found in the extracted features if the RNN is trained well. The model could classify malware processes with more preciseness by using a larger amount of data. Pascanu et al. experimented to learn the language of malware for the detection of unknown threats [44]. Bidirectional recurrent models were trained to predict next API call and use the hidden state that encodes the past event history as the fixed-length feature vectors. This is given to a separate classifier for the classification process. Max-Pooling is used over the values of the hidden units in the time since the hidden units may learn to specialize in detecting different and potentially reordered temporal patterns.

The efficacy of feature extraction using deep learning is utilized in the paradigm of anomaly detection in gas turbine combustors, where Stacked denoising auto encoders are used for the learning the features from the sensor readings of exhaust gas turbine combustors [45]. The features captured with deep learning approach perform better in acquiring the relationship between all sensor measurements and the latent behavior of the combustor compared to manual feature engineering. The learned features were fed into a neural network for identifying the anomaly in the measurements. This increased the performance of the anomaly detection system considerable. Also, the use of SDAE makes the system more immune to the noise in the input. The same characteristics of SDAE made Yao Wang et al. use them for the identification of malicious JavaScript code in web pages on the Internet. The use of feature engineering using deep learning without human intervention increased the detection accuracy of the classifier remarkably [46]. Salama et al. used DBN to reduce the dimension of feature sets making them appropriate for intrusion detection. The hybrid intelligent system combining the advantages of deep belief network and support vector machine. The reduced data output is improved by the use of DBN along with back-propagation. The model has the BP-DBN structure composed of 2 RBM layers. The data is reduced from 41 to 13 features by the first RBM layer and from 13 features to 5 output features by the second RBM layer on NSL-KDD data. DBN gives better performance than the other reduction methods (Table 5).

Table 5 Deep learning approaches for feature extraction

Sl No	Approaches for feature extraction	Inferences
1.	Feature extraction using stacked denoising auto encoders [46]	Extract more abstract features of JavaScript code
		Yields high classification accuracy compared to its shallow counterparts
2.	Auto Encoder for Dimensionality Reduction [47]	Space mapping ability of AutoEncoder's is utilized for reducing dimensionality of the data thereby abstracting the main characteristics.
		Restore data based on less information loss and error.
3.	Feature extraction and selection using auto encoders [26]	Reduce the manual work since the model is trained automatically once inputs of the model and stopping criterion of the iteration are determined.
4.	Feature extraction using stacked denoising auto encoders [45]	Features are explicitly learned without class labels
5.	Feature learning using Recurrent Neural Networks [48]	Language of malware is learned for the detection of unknown threats.
		Bidirectional recurrent models are used.
		Max-pooling is used over the values of the hidden units in time since the hidden units may learn to specialize in detecting different and potentially reordered temporal patterns.
6.	Recurrent Neural Networks for feature Extraction [43]	RNN is used to the extract features of the process behavior in a terminal.
		Trained features are converted to a feature image which is sent to classifier to be labeled malignant or benign.
		Regularity will be found in the extracted features RNN is trained well
7.	Deep belief networks for feature reduction [49].	Features were reduced considerably, from 41 features to 5 features in NSL KDD data
		DBN gives better performance than the other reduction methods.

2.4 *Deep Learning for Malware Detection*

Malware detection has emerged in the past years due to the rise in the threat caused by malware to large organizations. The major approaches for malware detection are static analysis and dynamic analysis [50]. The malware file or the group of files is evaluated precisely in the binary form or unpacked in the static analysis while, the binary files are executed, and the actions are reported in dynamic analysis. Dynamic detection is less exposed to obfuscation can offer direct observation of malware action, and makes it difficult to recycle existing malware. Static analysis, is exposed to obfuscation, and need no special set up for the data collection, but they are cooperative with the deep learning. Deep models perform efficiently in terms of number of fitting parameters than shallow networks. Table 6 summarizes the major works that uses deep learning for malware detection.

Saxe et al. proposed a malware detection approach based on deep neural networks (DNN) which achieves high a detection rate of 95% and a low false positive rate of 0.1% on an empirical dataset of over 400,000 software binaries [50]. This approach requires simple computation to perform feature extraction and it can achieve good accuracy. Even though the approach gives remarkable results, the performance collapse significantly in the time split validation since relying on syntactic features. Lie et al. has proposed a model which adapt to the environment to obtain remarkable detection of malicious code. They used DBN as a classifier for several times deep learning detecting malicious code. The detection accuracy was improved as the number of iterations was increased. The use of multiple deep learning shows better performance than surface learning model. An ensemble of deep feed forward networks and deep recurrent neural networks was used by Jung et al. for the detection of zero day flash malware detection [51]. Based on process behavior in possible infected terminals, Tobiyama et al. utilized the efficacy of CNN to annotate the

Table 6 Deep Learning Approaches for Malware detection

Sl No	Approaches for malware detection	Inferences
1.	Deep neural networks for malware detection [50].	Requires simple computation to perform feature extraction and it can achieve good accuracy Performance decays significantly in the time split validation since relying on syntactic features
2.	Malicious Code Detection using Deep Belief Networks [47]	Increase in the number of iterations in the DBN, increases the performance
3.	Malware Detection with CNN-RNN using Process Behavior [43]	Based on process behavior in possible infected terminals. CNN classify the malware process from the features extracted by RNN
4.	Visualized Malware Classification Based-on Convolutional Neural Network [52]	Malware features are converted to images and these features are fed into CNN for classification

behavior as malware or benign. Another deep learning technique RNN was used to convert the features collected to feature images such that they can be fed into CNN for the classification. Better performance was obtained for the work done by the authors of [52] by using the same approach.

3 Challenges and the Road Ahead

3.1 Challenges in Applying Deep Learning in IoT Security

The security issues in the Internet of Things are application specific, so are their solutions. With heterogeneous application contexts and various security requirements, they demand application-specific solutions. The network has to be equipped with technologies which can adapt to real time changes during the production or to foresee and refrain from events that might annihilate various operations. The IoT arena demands cross-layer security architecture since a quick fix solution is not applicable. Lightweight solutions that meet the specific requirements are to be designed for the specific application. The resource constraints and the limited computational capabilities of edge nodes are the major challenges for developing deep learning solutions in IoT.

3.2 Future Perspectives

3.2.1 Resource Constraint Deep Learning for Edge Computing

Adopting Artificial Intelligence and machine learning to the security of IoT, leveraging the efficiency of deep learning, reaps the reward of enhanced security in the system. Deep learning contributes to a feasible solution for the security scenario in the IoT networks to prevent the intrusions before any harm is done to the whole system. Deep-learning based algorithms surpass the explicit hand-made feature extraction methods amassed with traditional classifiers and can achieve equivalent accuracies for both noise-free and noisy data. The absence of manual feature manipulation, unsupervised pre-training and compression capabilities makes the application of deep learning beneficial for the resource constrained networks. Despite the distinguished performance of deep learning techniques, due to the increased computational complexity, there is a high demand in the designing of light weight versions of these techniques to make them resource friendly in the IoT scenario. Resource constrained solutions can be embedded in the device. The data processing needs can be contented "at the edge," where the data is collected, or where the user performs certain actions. Including additional capability of intelligence to process the data at the edge reduces the overhead of transmission of large chunks of data in real time. Furthermore, it reduces the response time to events by forbidding the transit of data to and fro the cloud for the computational purpose. Adding decision making capability closer to the devices contributes to the overall performance of the systems.

3.2.2 Adversarial Deep Learning

Adversarial deep learning has caught attention recently since they have evolved as a serious threat to the machine learning systems. It can be considered as a rendezvous of machine learning systems and cyber security systems. ADL which was concentrated in computer vision has disseminated to other domains too. ADL refers to the alteration in the original data to confuse the machine learning model and force them to misclassify the data. DL systems have to deal with mainly two types of attacks, Evasion, and poisoning. In evasion, the attacker alters the inherent behavior of the data to stay anonymous, and poisoning means the training data itself is altered. Robust solutions against the AML should be used along with intrusion detection systems to make the approaches impervious. Adversary samples are made with the help of evolutionary algorithms, Fast Gradient Sign method (FGSM) and Jacobian-based Saliency Map Attack (JSMA). The variations done are hard to be sensed by the humans. Although solutions like distillation, incorporation of the adversarial component in the loss function, training with adversarial samples first to reduce the effect, etc. have been proposed, there exists a large realm for the researchers interested in ADL to work on.

4 Conclusion

The revolution of connectivity that brewed around the world in the past three decades gave rise to the third wave in the development of internet, Internet of Things which became an inevitable part of human life. Big data analytics harness the massive amount of data generated by Internet of Things and convert to well-analyzed data which is extremely valuable in today's world. To discover the sophisticated latent features abstract deep learning techniques are used. This abstraction ability and capability to handle the enormous amount of data tactfully, makes it a major focus in data science. In this chapter we have tried to limelight different deep learning approaches utilized in the area of cyber security. It gives a broad analysis of the deep learning techniques for feature extraction and classification tasks like intrusion detection, malware analysis, authentication etc. This chapter provides a sketch of the state of the art deep learning techniques, challenges faced and pointers to the future research direction. With Internet of things as the senses, big data as the powering force and the deep learning as central processing pivot, we can realize a smart connected world in future.

References

1. George G, Thampi SM (2018) A graph-based security framework for securing industrial IoT networks from vulnerability exploitations. IEEE Access 6(September):43586–43601
2. Sicari S, Rizzardi A, Grieco LA, Coen-porisini A (2015) Security, privacy and trust in internet of things: the road ahead. Comput Netw 76:146–164
3. Jing Q, Vasilakos AV, Wan J, Lu J, Qiu D (2014) Security of the Internet of Things: perspectives and challenges. Wirel Netw 20(8):2481–2501
4. Lecun Y, Bengio Y, Hinton G (2015) Deep learning. Nature 521(7553):436–444
5. Kim, Kwangju, Muhamad Erza Aminanto, and Harry Chandra Tanuwidjaja.(2018) Network Intrusion Detection Using Deep Learning: A Feature Learning Approach. Springer.
6. Aminanto, M. E., & Kim, K. (2016) Deep learning-based feature selection for intrusion detection system in transport layer. In Proceedings of the Summer Conference of Korea Information Security Society (CISC-S'16), pp 535–538, 2016
7. Rooshenas A, Lowd D (2014) Learning sum-product networks with direct and indirect variable interactions. Proc 31st Int Conf Mach Learn 32:710–718
8. Poon, H., & Domingos, P. (2011, November). Sum-product networks: A new deep architecture. In 2011 IEEE International Conference on Computer Vision Workshops (ICCV Workshops) (pp. 689–690). IEEE.
9. Kim, J. W. Classification with Deep Belief Networks." https://www.ki.tu-berlin.de/fileadmin/fg135/publikationen/Hebbo_2013_CDB.pdf
10. A. Elsherif(2018) "Automatic intrusion detection system using deep recurrent neural network paradigm," Journal of Information Security and Cybercrimes Research (JISCR), vol. 1, no. 1, 2018.
11. J. P.-A. Ian J. Goodfellow, D.-F. , Mehdi Mirza, Bing Xu, S. Ozair†, and Y. B., Aaron Courville, "Generative Adversarial Nets," *arXhiv*, vol. 155, no. 4, pp. 270–275, 2013
12. Saleema, A., & Thampi, S. M. (2018) Voice Biometrics: The Promising Future of Authentication in the Internet of Things. In Handbook of Research on Cloud and Fog Computing Infrastructures for Data Science (pp. 360–389). IGI Global.
13. Lee YS et al (2016) Touch based active user authentication using deep belief networks and random forests. Proc 6th Int Conf Inf Commun Manag ICICM 2016:304–308
14. Maheshwary S, Ganguly S, Pudi V (2017) Deep secure: a fast and simple neural network based approach for user authentication and identification via keystroke dynamics. IWAISe First Int Work Artif Intell Secur 2017:59
15. Shi C, Liu J, Liu H, Chen Y (2017) Smart user authentication through actuation of daily activities leveraging WiFi-enabled IoT. In: Proceedings of the 18th ACM international symposium on mobile Ad Hoc networking and computing – Mobihoc'17, pp 1–10
16. Das R, Gadre A, Zhang S, Kumar S, Moura JMF (2018) A deep learning approach to IoT authentication. In: IEEE international conference communication, vol. 2018–May
17. A. Ferdowsi and W. Saad (2018) Deep learning for signal authentication and security in massive Internet of Things systems, pp 1–30
18. Rajasegarar S, Leckie C, Palaniswami M (2008) Anomaly detection in wireless sensor networks. IEEE Wirel Commun
19. Yavuz, F. Y. (2018) Deep learning in cybersecurity for internet of things (Doctoral dissertation).
20. Diro AA, Chilamkurti N (2018) Distributed attack detection scheme using deep learning approach for internet of things. Futur Gener Comput Syst 82:761–768
21. Kim J, Shin N, Jo SY, Kim SH (2017) Method of Intrusion detection using deep neural network. Int Conf Big Data Smart Comput:313–316
22. Ma T, Wang F, Cheng J, Yu Y, Chen X (2016) A hybrid spectral clustering and deep neural network ensemble algorithm for intrusion detection in sensor networks. Sensors 16(10):1701
23. Kang M, Kang J (2016) Neural network for in-vehicle network security. PLOS One 11:1–17
24. Li, Y., Ma, R., & Jiao, R. (2015) A hybrid malicious code detection method based on deep learning. International Journal of Security and Its Applications, 9(5), 205–216.

25. Niyaz Q, Sun W, Javaid AY, Alam M (2015) A deep learning approach for network intrusion detection system. In: Proceedings of 9th EAI international conference Bio-inspired Information and Communication Technologies
26. Wang Z (2015) The applications of deep learning on traffic identification. Black Hat, Washington, DC
27. Alom MZ, Bontupalli V, Taha TM (2015) Intrusion detection using deep belief networks. In: 2015 National Aerospace & Electronics Conference, pp 339–344
28. Gao N, Gao L, Gao Q, Wang H (2015) An intrusion detection model based on deep belief networks. In: Proceedings – 2014 2nd international conference on advanced Cloud Big Data, CBD 2014, pp 247–252
29. Fiore U, Palmieri F, Castiglione A, De Santis A (2013) Neurocomputing Network anomaly detection with the restricted Boltzmann machine. Neurocomputing:1–11
30. Fiore U, Palmieri F, Castiglione A, De Santis A (2013) Network anomaly detection with the restricted Boltzmann machine. Neurocomputing 122:13–23
31. Dong B, Wang X (2016) Comparison deep learning method to traditional methods using for network intrusion detection. In: 8th IEEE international conference on communication software networks, pp 581–585
32. Sheikhan M, Jadidi Z, Farrokhi A (2012) Intrusion detection using reduced-size RNN based on feature grouping. Neural Comput Appl 21(6):1185–1190
33. Chuan-long Y, Yue-fei Z, Jin-long F, Xin-zheng H (2017) A deep learning approach for Intrusion detection using recurrent neural networks. IEEE Access 5:1–1
34. Lopez-Martin M, Carro B, Sanchez-Esguevillas A, Lloret J (2017) Network traffic classifier with convolutional and recurrent neural networks for internet of things. IEEE Access 5:18042–18050
35. Kim J, Kim J, Thu HLT, Kim H (2016) Long short term memory recurrent neural network classifier for intrusion detection. In: 2016 international conference on platform technology and service, no. February, pp 1–5
36. Bediako PK (2017) Long short-term memory recurrent neural network for detecting DDoS flooding attacks within TensorFlow Implementation framework
37. Kim G, Yi H, Lee J, Paek Y, Yoon S (2017) LSTM-based system-call language modeling and ensemble method for host-based intrusion detection. pp 1–12
38. Cheng M, Li Q, Lv J, Liu W, Wang J (2018) Multi-scale LSTM model for BGP anomaly classification. IEEE Trans Serv Comput, no NetworkML:1–6
39. Putchala MK (2017) Deep learning approach for Intrusion Detection System (IDS) in the Internet of Things (IoT) network using Gated Recurrent neural networks (GRU) p 63
40. Cheng M, Xu Q, Lv J, Liu W, Li Q, Wang J (2016) MS-LSTM: a multi-scale LSTM model for BGP anomaly detection, no. NetworkML, pp 1–6
41. Lopez-martin M, Member S, Carro B (2017) Network traffic classifier with convolutional and recurrent neural networks for Internet of Things. IEE Access 5
42. Vinayakumar R, Kp S, Poornachandran P (2017) Applying convolutional neural Network for Network Intrusion detection, pp 1222–1228
43. Tobiyama S, Yamaguchi Y, Shimada H, Ikuse T, Yagi T (2016) Malware detection with deep neural network using process behavior. In: 2016 IEEE 40th annual computer software and applications conference, pp 577–582
44. R. Pascanu, M. Marinescu, and A. Thomas (2015) Malware classification with recurrent networks. In: IEEE international conference on Acoustics, Speech and Signal Processing – Proceedings, v2015-August, pp 1916–1920
45. Yan W, Yu L (2015) On accurate and reliable anomaly detection for gas turbine combustors : a deep learning approach. PHM Conf:1–8
46. W. C. and P. W. Yao Wang∗ (2016) A deep learning approach for detecting malicious JavaScript code. Secur Commun NETWORKS Secur Comm Networks 2016, 9(22):1520–1534
47. Li Y, Ma R, Jiao R (2015) A hybrid malicious code detection method based on deep learning. Int J Secur Its Appl 9(5):205–216

48. Pascanu R, Stokes JW, Sanossian H, Marinescu M, Thomas A (2015) Malware classification with recurrent networks. In: ICASSP, IEEE internation conference Acoustics speech signal process. – Proceedings, vol. 2015–August, pp 1916–1920
49. Salama MA, Eid HF, Ramadan RA, Darwish A. 103_Hybrid Intelligent Intrusion Detection Scheme.pdf, pp 1–11
50. Saxe J, Berlin K (2015) Deep neural network based malware detection using two dimensional binary program features
51. Jung W, Kim S (2015) Poster: deep learning for zero-day flash malware detection. Proc IEEE Symp Secur Priv:2–3
52. Seok S, Kim H (2016) Visualized malware classification based-on convolutional neural network. J Korea Inst Inf Secur Cryptol 26(1):197–208

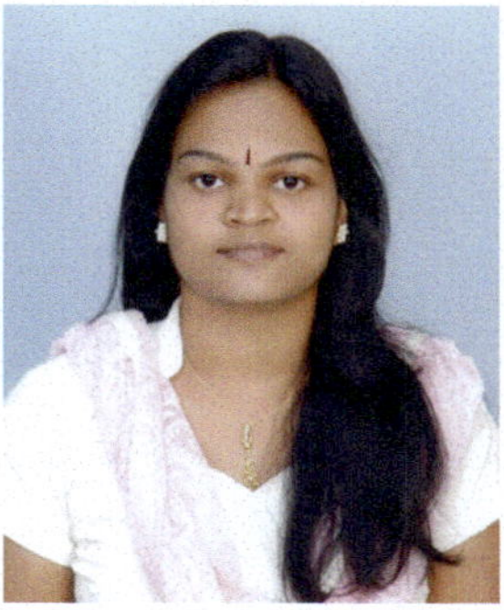

K. S. Sunitha Krishnan received her M.Tech degree in Computer Science, with specialization in Software Engineering, from the Department of Computer Science, Cochin University of Science and Technology in 2011 and B.Tech degree in Computer Science and Engineering from Mahatma Gandhi University in 2004. Her research interests include Machine Learning, Cybersecurity, Intrusion Detection, data analytics and Internet of Things (IoT).

Sabu M. Thampi is a Professor at the Indian Institute of Information Technology and Management, Kerala (IIITM-K), Trivandrum, India. He has completed his Ph.D in computer engineering from the National Institute of Technology, Karnataka. His research interests include network security, security informatics, bio-inspired computing, video surveillance, cloud security, secure information sharing, secure localization, and distributed computing. He has authored and edited few books published by reputed international publishers and published papers in academic journals and international and national proceedings. He is currently serving as Editor for Journal of Network and Computer Applications (JNCA), Elsevier and Journal of Applied Soft Computing, Elsevier; and Associate Editor for IEEE Access and International Journal of Embedded Systems, Inderscience, UK; and reviewer for several reputed international journals. He is a Senior Member of IEEE and member of IEEE Communications Society, IEEE SMCS, and ACM.

Deep Learning Meets Malware Detection: An Investigation

Biozid Bostami and Mohiuddin Ahmed

Abstract From the dawn of computer programs, malware programs were originated and still with us. With evolving of technology, malware programs are also evolving. It is considered as one of the prime issues regarding cyber world security. Damage caused by the malware programs ranges from system failure to financial loss. Traditional approach for malware classification approach are not very suitable for advance malware programs. For the continuously evolving malware ecosystem deep learning approaches are more suitable as they are faster and can predict malware more effectively. To our best of knowledge, there has not substantial research done on deep learning based malware detection on different sectors like: IoT, Biomedical sectors and Cloud platforms. The key contribution of this chapter will be creating directions of malware detection depending on deep learning. The chapter will be beneficial for graduate level students, academicians and researchers in this application domain.

1 Introduction

As technology is evolving at high speed, malware programs are becoming smarter than before. The malware programs are still considered to be main threat to the cyber security. By definition, malware programs are computer programs which are designed to impose harm to user's computer in various ways e.g.: stealing data, destroying data etc. As the malware programs are changing rapidly, the anti-virus scanners are failing to provide the protection. As a result more systems are becoming victim of malware attack everyday. According to the report published by kaspersky in 2017 around 174,989,956 new malware are being discovered. Also many attacking tools are now available through internet. Malware programs can cause

B. Bostami (✉)
Islamic University of Technology, Dhaka, Bangladesh

M. Ahmed
Lecturer of Computing and Security, School of Science, Academic Centre of Cyber Security Excellence (ACCSE), Edith Cowan University, Joondalup, WA, Australia
e-mail: m.ahmed.au@ieee.org

Z. Md. Fadlullah, A.-S. Khan Pathan (eds.), *Combating Security Challenges in the Age of Big Data*, Advanced Sciences and Technologies for Security Applications, https://doi.org/10.1007/978-3-030-35642-2_7

major data damage even with a single attack. Android based are also being targeted by cyber criminals for exploiting sensitive and personal data. These platforms have different architecture that is why the malware programs are also modified. Malware programs can even spread through hybrid platforms e.g.: a windows based computer connected to other linux based server can affect other linux servers with malware and vice versa. Some malware might not harm the user data or system but there are some malware like ransomware can do a huge economical loss. Researchers have been working on malware classification and detection and presented a many different approaches over the past years. Classical malware detection methods were based upon mainly signature based or behaviour based. But with the discovery of polymorphic malware programs with the ability to change the code classical approach could not detect the new malware properly. Then machine learning approaches were proposed by the researchers. The machine learning also had some limitations but the methods showed higher accuracy over the classical approach and able to detect the unknown malware programs with mutation property. With the advent of Deep learning the malware detection and classification techniques also raised up to a new level. Deep learning approaches has more accuracy over the machine learning methods. Also, they are faster than machine learning algorithm in detecting malware. In this chapter we mainly explored different methods of malware detection. Additionally, we will put light to the malware detection on Android malware sectors which need to addressed by the research community.

2 Chapter RoadMap

The following chapter is organized as follows: We start by introducing malware programs and classifications. Then we revisited the classical detection technologies followed by the classical machine learning approaches. After that, we reviews the deep learning approach in malware detection. Then we focus on present state of android malware detection. Finally conclude the chapter with conclusion.

3 Malware

The word malware is originated from two words: malevolent and software. It is also used to indicate unwanted and harmful software or computer program. According to G. McGraw and G. Morrisett who defines malware programs by following: "any code added, changed, or removed from a software system in order to intentionally cause harm or subvert the intended function of the system" [1]. Xufang et al. [2] defines malware programs as worms, rootkits, trojans and other malicious intrusive programs. In 1983 F. Cohen defines the well known malware known as 'virus' to be a computer program which infect and alter other programs in order to attach an evolved copy of itself [3].

Malware programs are unwanted softwares. Due to the infection ability, the malware can propagate through the computer networks. Other abilities and characteristics of malware programs include camouflage, self-replication, self-execution, and manipulating other programs. Self-replication is a common characteristics for all malware programs which ensures their existence in the cyber world. The camouflage ability of the malware helps them to remain undetected in the system. Another important and interesting ability of the malware programs is self-execution. Most the malware do not need any user interaction to execute itself. Once it is initiated these malware programs can execute independently. The main harmful nature of malware programs is that they are not harmless. Very few malware may seemed harmless but most of the malware programs are created with the intention of corrupting the system in which it resides.

Some malware programs may alter valuable data, some malware programs are used to gain monetary gains, even some malware programs disrupts the systems incurring huge financial losses.

The usual way of malware infection is to transfer of malicious programs to other systems. This transfer can be done in many ways e.g. the user can install the malware program directly or the malware programs can themselves affect other connected devices by exploiting security. Once a malware is initiated, it can also use the captured device to attack other devices disrupting the normal flow of operation.

4 Types of Malware

There are wide ranges of malware programs in the cyber world. With the aid of advance technology more malware programs are being created by the cyber criminals. But the behaviour and purpose of creation remain the same. The primary purpose of these classification is to correlate the different information to further identifying unknown and new malware [4]. Based on the network and propagation mechanism malware can be classified into two major classes: (1) Ordinary Malware, (2) Network-Based Malware.

5 Ordinary Malware

These types of malware resides in the local file systems. They normally use the media storages to propagate from one machine to another machine. Most of these types can be easily detected based on the signature-based detectors.

Viruses belongs to this family of malware programs. Viruses are self replicating program with an urge to reproduce itself based on how it was programmed. During the infection stage it can infect other files and applications. Once it is initiated viruses can spread rapidly infecting the system it resides which may cause to system failure. Viruses are attached to different application following methods like:

concatenation and embedding techniques [5, 6]. For example: cyber criminals attach their malicious code into the 'Autorun.inf', a file that is searched by the operating system when a removal device is attached.

Similar to computer viruses worms has the ability to replicate itself in the infected device. They cause similar damages to the system as viruses. But, worms are independent program. It can complete own life cycle without another software. It can propagate without human interaction or any other program/software. To spread, worms exploit the system vulnerabilities or trick the users through social engineering to execute itself. Once initiated it could use the system features to travel undetected causing harm to the system [7]. Worms can prevent the legitimate users by consuming bandwidth and system resources.

A malware named as 'Logic Bomb' executes when all the logical conditions are met. Among the common activation factor are date and time. It continuously checks for system date and time for deciding whether to execute or not. When the condition of date and time meet the malware executes [5].

6 Network-Based Malware

In this category Trojans, Spywares, Adwares, BackDoor, Rootkit, Spam, RansomWare, Cookies etc. are well known. Among these malware programs one exception is Cookies. Unlike other malware cookie itself is not an executable program. These are plain text files. They hold device information by the web browser. The primary purpose of cookie is to hold server side sessions and user authentication factors for different sites. Cookies are send to the user by the web server and sent back with each HTTP request. Cookies may be permanent or they may expire after certain time period. Cookies do not harm the system directly but cookies can be used by other malware programs like spywares to exploit the system. That is why it is also mentioned in the list.

Spywares are softwares which are installed by other valid softwares without user's direct permission. The spywares comes along with a valid packages of software, tricking the users to install it on their system all together. These spyware collects user information from the device and sent back to different servers. Some reputed vendors like Microsoft, Google etc. secretly collects user information using spywares [5, 8].

Adwares are softwares created for generating revenue based on auto-generating advertisements on other software infaces or during installation steps [7]. Most of the Adwares are safe by nature. But it cause interrupt users by showing unnecessary ads on the screen blocking the main view. Then again not all the Adwares harmless, some comes with a integrated key-logger or other spywares along with it [9].

BackDoor, grant attackers the full access to system bypassing the normal authentication. Backdoors are normally created by the primary developers of the system or set by other programmers through compromising the system [7]. Attackers use backdoors for carrying out future attacks. Sniffers are special kind of malware that intercept the network traffics and collect valuable data. These collected data are then analyzed to carry out intrusion attacks.

Trojans are a kind of malware which seemed to be an legitimate software to users but actually harmful. The name ‘trojan’ was originated from the wooden horse of Greeks history which was used to invade troy. Once a trojan malware is activated into a system, it can do damages ranges from disrupting the user works by showing pop-up windows frequently to damaging the files and even placing backdoors for other malware like: viruses and worms [7, 9]. Unlike other malware trojan do not self-replicate or reproduce. It spreads through using sending emails to the users attaching itself or by tricking users to get it downloaded from the internet.

Ransomwares are malevolent program that blocks, encrypt the users stored informations or threatens to publish those and demands ransom [7]. Even after payment of ransom the retrieval of data is not ensured. Some ransomware blocks the root access which makes it impossible to remove. Many company faces a huge financial loss due to ransomwares. Some of the well known ransomware are: Cryptowall, CryptoLocker.

Bots are programs to automate tasks. The word is originated from ‘Robot’. These programs can be used for good or bad. Malicious bots are self-propagating. They can be designed to carry out attacks like Denial of Service attacks. Some bots collects financial information from the compromised systems and send to attacker’s original server. Some malicious bots can even place backdoors to the compromised system allowing the other malware programs to exploit the network [7]. Advance malicious bot can be carry out attacks like: false data injection attacks in Internet of Things (IoT) [10]. Crypto Mining is a common example of how malicious bots can be used [7].

Another Type of malware programs are known as mobile code [7]. These malicious codes are transmitted through remote servers to local servers and then they execute in local server. These codes consist of java, ActiveX or Jsx Script etc.

Malware creators always try to improve their malicious programs so that they are not easily detected by the detectors and get enough time and scope spread more. That is why they use different methods. Based on the technology used for creating and evading detectors malware programs are classified into four classes: (1) Encrypted Malware, (2) Oligomorphic Malware, (3) PolyMorphic Malware and (4) Metamorphic Malware.

7 Encrypted Malware

The simplest method of hiding the malicious code to evade analyzer is encryption techniques. The first encrypted malware was created named Cascade [11] back in 1987. These malware programs can be divided into two parts: (1) main encryption body, (2) Decryption Code. Main malicious code is made encrypted by combination of different keys making different signatures of the same malware. The encrypted part is the contains the malicious code which remain meaningless at beginning. Th

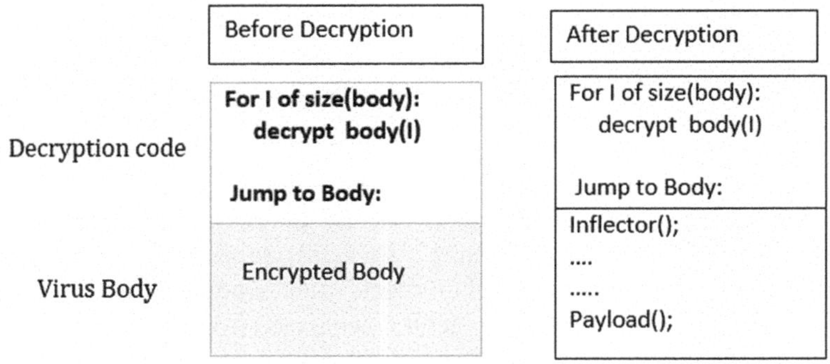

Fig. 1 Basic structure of encrypted malware

decryption code is responsible for producing the actual malware from the encrypted body once it is executed in the victim's machine. Encryption malware programs were discovered to evade code temparing, statical analysis. The main drawback is that the signature based detectors can detect these malware based on the decryptor part signature (Fig. 1).

8 Oligomorphic Malware

In order to overcome the drawbacks of encrypted malware, new malware programs were created where the decryptor section mutate with each infection. These malware are also called semi-polymorphic malware. These malware use set of different decryptor part by which each instance of malware poses different signature. The first oligomorphic malware known as 'Whale' was documented in 1990 [12]. Win95/Memorial, an oligomorphic malware had about 96 different decryptor signatures which proofs that detection based on decryptors alone is not accurate [13] (Fig. 2).

9 Polymorphic Malware

Polymorphic malware poses the ability to different forms on its own. It can mutate decryptor section in the range of millions [12]. First documented polymorphic malware was found in 1990 known as '1260 virus' created by Mark Washburn [12]. These malware uses code obfuscation methods like substitution or addition of junk code to mutate the decryptors in each instance [2]. The section which controls the mutation is known as mutation engine. Two advance polymorphic malware were HPS and Marburg, founded by GriYo in 1998. The mutation engine of HPS was relatively advance compared to Marburg. The detailed information can be found in [12, 14]. The main drawback of polymorphic malware is that they eventually decrypt

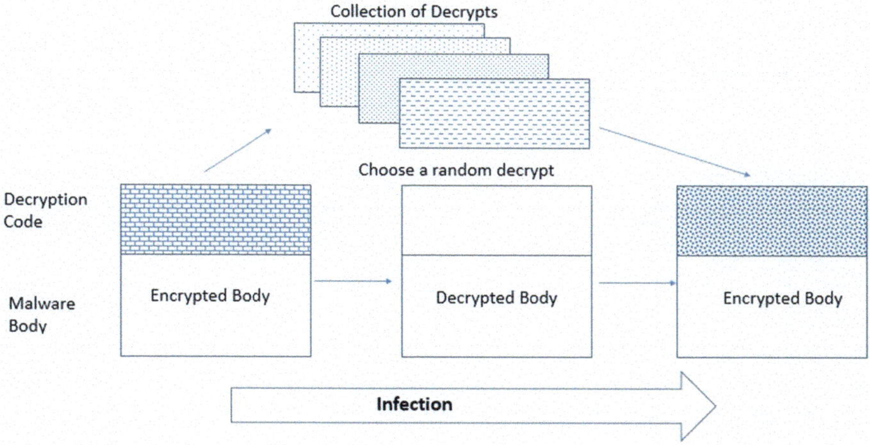

Fig. 2 Basic structure of oligomorphic malware

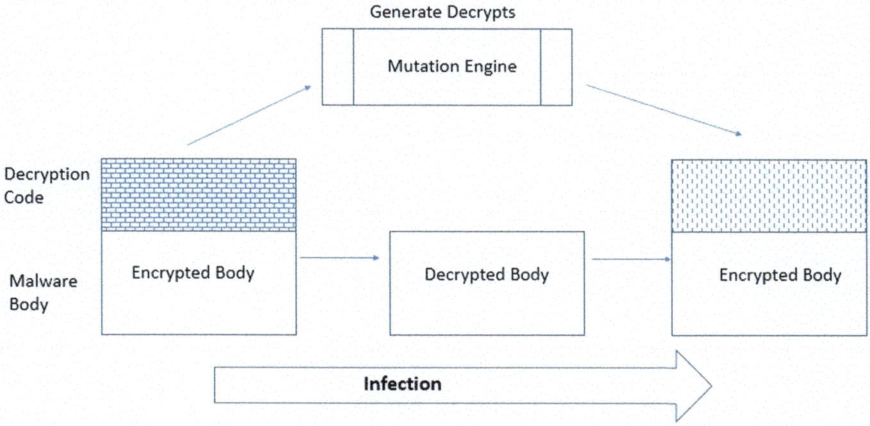

Fig. 3 Basic structure of polymorphic malware

itself to its original form. The advance detection algorithm thus easily detect them easily. Another drawback is the maintenance and creating polymorphic malware is much difficult and time consuming [15] (Fig. 3).

10 Metamorphic Malware

Metamorphic malware programs are quite different from polymorphic and oligomorphic malware. Unlike others it do not have encryptor part as well as it has no decryptor part as well. But it has a mutation engine. Igor Muttik mentioned simple

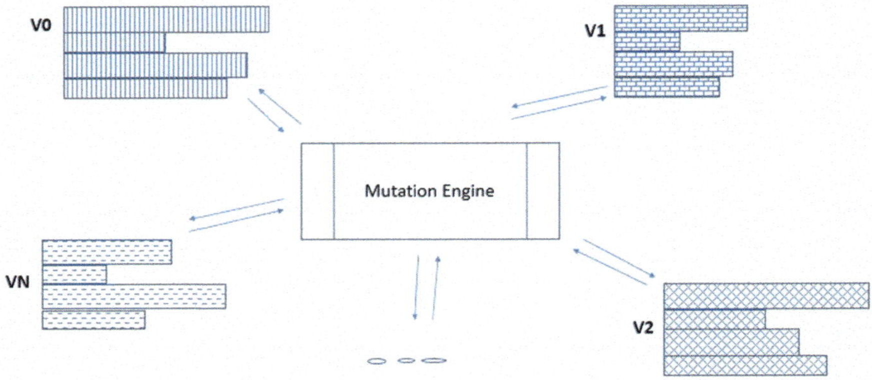

Fig. 4 Metamorphic malware propagation method

and precise definition of metamorphic malware as: "Metamorphics are body-polymorphics" [12]. Even though every instance of the metamorphic malware programs consist of different size, property, code sequence, size, structure but the behaviour remain same. Metamorphic malware avoids creating variant similar to their ancestor. Metamorphic malware uses different methods for creating new unique instances. Such methods are: garbage code insertion, register usage exchange, insertion of jump sequence, code integration, code mutation etc. [16, 17]. Creating a metamorphic malware which retain its size fixed is a challenging task. The first metamorphic malware documented was win95/Regswap created at 1998 by [18]. Metamorphic malwares changes its own code using the mutation engine. Some of advanced metamorphic malware programs documented are: Win95/Zmist, Win95/Bistro, Win32/Ghost and the Win95/Zperm (Fig. 4).

11 Classical Malware Detection Methods

Ross Greenberg founded the first anti-malware software was documented in 1987 named 'Flushot Plus'. It could prevent trojan and viruses to make unauthorized alteration to other softwares. John McAfee created the VirusScan program in 1989. This anti-malware could repair and detect virus files on windows. After that many malware classical detection methods were proposed by different researchers. These classical approach are classified into three main category: (1) Signature Based Detection, (2) Behaviour Based Approach and (3) Specification Based Detection.

12 Signature Based Detection

Malware programs can be detected comparing their content signature with some known malware pattern. These pattern matching based methods are common in malware detection field [19]. Signatures of different malware programs are collected and preserved by the anti-virus companies. The signature database is refreshed periodically. The malware signatures contains unique identification data about a specific malware family. The details of signature based detection and analysis of malware can be found in details in [20]. The main drawback of this signature based detection is collecting and maintaining of signatures of malware programs. As, metamorphic malware produce different and unique signatures for each instance, signature based detection fails to detect these classes of software. Moreover, to build signature of a malware consumes a lot of time.

13 Behaviour-Based Approach

In these detection methods malware programs are identified based on how they behave in certain environments [21]. In these detection techniques the programs with similarity in their behaviour are collected and based on the behaviour signature a program is labelled as malicious or safe. Behaviour based detection can easily detect the malware which can mutate their code signatures but have same behaviour patterns e.g. the system calls, services and resource allocations etc. The tasks of behaviour based detector can be divided into three stages [22] as following:

1. **Data Collector**: In the first stage behavioural data is collected from from the malware executables in static or dynamic nature.
2. **Interpreter:** In this stage the collected data by the collector is transformed into intermediate representable format.
3. **Matcher:** In the final stage the matcher test the formatted information with the known signatures to make the decisions.

Symantec proposed histogram based malware detection which is an example of behaviour based detection. Hofmeyr et al. [23] had consider system call sequence to detect malware programs. Sato et al. [24] consider frequency distribution of the system calls by malware programs. Though these approaches can detect different malware programs with multiple code signature very efficiently but it is very scan time consumption and false positive ratio is quite high, which are the main disadvantage.

14 Specification Based Detection

Specification based detection set some defined rules and based on those rules detect malware programs. It addresses the limitation of the behaviour based detection which is high false alarm rate. These algorithm tries to approximate the requirements of the application unlike behaviour based approach. In the training stage, algorithm is feed with some set of valid rules for a system which is inspected. The fundamental drawback in these approaches is that setting the rules of specification is difficult tasks to do when it is for complex system. Even some common specification become difficult to transfer into machine instruction. Again, setting the complete set of valid rules set is takes time and sometime is not even possible to specify all the valid rules. One example of the specification based detection tools is Panorama [25].

15 Machine Learning Based Malware Detection Revisited

The first trial for integrating machine learning approach was proposed by Schultz et al. [26]. They trained naive bayes and multi naive bayes algorithms to distinguish malware from normal files. Since machine learning algorithms decides based upon different features e.g. api or system calls, control flow graph etc. so different researchers proposed their work on different features.

Henchiri and Japkowicz in their work presented at [27] four machine learning classifier: ID3, J48, SVM and Naive Bayes for classification of malware. Their work shows low rate of false positive. Hofmeyr et al. in his work proposed the system call to detect malware [23]. In their proposal programs with short sequence of api call to system was considered as safe. They proposed hamming distance along with a threshold value to detect anomalies. Larger distant values were considered to anomalies.

Ye et al. presented Intelligent Malware Detection (IMDS) [28]. Their work they used Object Oriented Association (OOA) mining for classify malware. Their work was based on windows portable executable files. The OOA classification rules were created by OOA-FAST-FP algorithm. This system had two major drawbacks: (1) Building the proper rules for unseen malware. (2) Processing the larger rule sets for classifier.

In order to overcome the drawbacks Ye et al. [29] presented CIDCPF. In their work they apply Chi-squared testing and pessimistic error estimation on the database. Then they made the prediction fetching the best first rule. Then merge the CIDCPF and IMDS and merged system is named CIMDS [29]. This system is considered to be an initial attempt to incorporate the associative classifier in malware detection which uses the post processing. The only limitation is it gives only binary predictions.

Jeong and Lee in their work also consider the API call and presented code graph to detect malware [30]. They created code graph using the topological tree of system calls of executable programs. The code graph of the binary programs are tested against the malware code graph for classification. The main drawbacks were that the size of the code graph was very large. To overcome the limitations of his previous divided the system api calls into 128 categories in [31].

Ye et al. in presented the hierarchical associative classifier [32]. It is build upon the system API calls gathered from a grey list. The list was taken from King soft corporation. Ahmed et al. [33] in his work use API calls with the spatial and temporal feature collected from malware executables. They use five classifier to malware classification.

OpCode (Operational Code) is used to defines the set of operations which get executed when program runs. It is considered to be the subunit of machine instruction which composed of operational codes and operands. For malware detection this feature is also used by different researchers. Bilar et al. [34] in his research presented that OpCodes can be a feature to detect malware. Santos et al. also consider OpCode sequence to detect malware and presented some algorithm. One of their work considered the appearing frequency of the OpCodes to detect obfuscated malware programs [35]. They disassemble executable files and created OpCode profile then compute the appearance frequency of OpCodes for malware and safe datasets using mutual data. Then the feature vector was extracted from the files using weighted terms frequency. This feature vector was used to detect obfuscated malware programs.

Santos et al. in his future work presented different feature extraction algorithms depending on OpCode call sequence [36]. He trained different machine learning classifier with the extracted features. Santos et al. also presented different machine learning based approach with a view to overcoming the drawback that classifiers need large amount of labeled data to train which are harder to find in real world. His future researches presented single class learning [37] and collective Classifier [38]. Santos et al. [39] also presented semi-supervised classifier algorithm to detect malware.

Runwal et al. also presented methods based on OpCode for metamorphic malware detection [40]. In their work they use graph based matching for detection. The graph are created from the OpCodes from the malicious and benign files comparing the pairs of similar OpCodes. Then new files are tested against the OpCode graph for similarity for detection.

Shabtai et al. also presented a classifier based on OpCode patterns [41]. They collected the features such as: (1) Term Frequency (TF), (2) TF-Inverse Document frequency. Then train different machine learning classifier like: SVM, ANN, LR etc. Their work consider the windows executables file mainly.

Another feature for malware detection using machine learning is called N-Grams. By definition, N-Gram for a string is a set containing all substrings each of length N. For example: some 4-Grams of the string 'MALWARE' are: 'MALW', 'ALWA', 'LWAR', 'WARE' and so on. Being motivated, many researchers has presented new methods to detect unknown malware programs based on the binary contents patterns.

Schultz in his work [26] introduced the concept of malware detection from binary codes. He proposed different feature mining methods e.g.: plain text extraction, bytecode sequence, PE section.

Tesauro et al. [42] presented malware detection method where N-Gram was used as a feature. They targeted the boot sector virus, which hampering normal nature of the system as it damages the boot files. N-Grams were built from the malware and benign programs. They also used feature reduction methods to select least N-Gram to choose from collected N-Gram sets. They used Artificial Neural Networks (ANN) to detect the malware.

Tesauro et al. [43] in their future work presented several classifiers using ANN and result was collected using voting strategy. They also reduced the N-Grams based upon threshold value.

Authors in [44] proposed N-Gram with K-Nearest-Neighbour (KNN). Kotler and Maloof in their work [45] considered binary representation of N-Gram for detecting malware. Moskovitch et al. [46] considered the byte N-Gram where they illustrate the imbalance problem associated to the dataset.

Bruschi et al. [47] in their work normalized the executable and minimize the mutation effects and generate control flow graph (CFG). Then the malware CFG is compared with general CFG to check if it contains an isomorphic subgraph of the normalized CFG. Thus malware detection becomes sub-graph isomorphism problem.

Zaho et al. [48] presented techniques based on the extracting features from CFG to detect malware. These features include node, subgraph and edges. Data is trained with the extracted features from CFG. The data mining methods are used for malware classification e.g.: Decision tree and Random Forest.

Bonfante in his work presented in [49] used the CFG as malware signature for classification. As we know CFG contains 4 instruction e.g.: jump, conditional jump, function call and return. Along these instruction they introduced two more instruction node named: 'inst' for contiguous instruction sequence and 'end' for program end. After they initially build the CFG they reduced it following the rules: for a node 'inst' or 'jmp' the particular node is removed and connect its predecessors to the uniq successor. Thus reduced graph becomes the signature file for malware detection. The CFG based approach works for simple malware programs but can not perform on complex malware with mutation ability.

Eskandari in their work [50] presented a combination CFG and API call for metamorphic malware. They used API call to CFG to get better understanding of the malware semantic. To graph algorithm complexity was reduced by converting the graph into feature vector. Selection methods were proposed to selecting features and train the sample data to train, producing the rule dataset. Then decision system make the final decision based on the rule set.

Kim and Moon [51] presented dependency graph approach to detect the polymorphic malware. They targeted the script viruses. First the malware are transformed into semantic code, then the codes are transformed into directed graph. Following some rules the directed graph is transformed into dependency graph. For unknown malware graphs are compared and thus the malware detection is done by solving maximum subgraph isomorphism.

Nataraj in his work presented at [52] represent byte code of executables into grayscale image. They transform the malware detection into image classifier problem. They wrap the malware image data into 2-D matrix form and apply different feature extraction methods and machine learning classifier to determine malware.

These machine learning based detection poses a common drawback which is they all heavily depend on expert knowledge of feature extraction design. While the malware prograams are continuously changing and evolving and adding or modifying their features, machine learning algorithm can not keep up the pace as updating the human designed feature selection takes a significant amount of time.

16 Deep Learning Based Detection

Deep learning is a advance form of machine learning algorithms. For addressing the limitations faced by classical machine learning approaches for detecting malware many researchers proposed some deep learning algorithm. Deep learning algorithm can work with large number of features and can more accurate than machine learning algorithms. Here we presented some approaches for detecting malware based on deep learning proposed by the researchers.

Dhal et al. [53] presented deep learning using random projection and neural network. In their work they also presented that increasing the hidden layers do not affect the accuracy much. Saxe and Berlin [54] worked with feedforward based neural network for malware detection. It was a static approach. No dynamic classification was found in their work. Pascanu et al. [55] proposed recurrent network to model the system call sequencing for building a corpus for malware.

Cakir et al. [56] in their work presented deep learning using Word2Vec for malware presentation and gradient search to classify malware. Their work shows high accuracy over detecting unknown malware. Raff et al. [57] proposed a static malware analysis using binary information form the application. The used raw byte from the programs and build a neural network to train and decide. It can detect malware without running the program itself.

David et al. [58] introduced signature based malware detection using deep learning. The malware behaviour were analyzed based using the deep belief network (DFN). Their method showed 98.6% accuracy over malware dataset.

Elmouatez et al. [59] proposed MalDozer which is tool for detecting malware programs in android system. The tool uses deep learning with raw API calls.

Sunho et al. [60] presented malware detection method based on deep learning where they used malware images to along with CNN to classification. First the malware images are extracted and then train CNN with the images. New malicious programs are classified based on the image comparison. Their report shows 96% accuracy.

Boydell et al. [61] in their work consider binary files to represent the malware using 1-dimensional binary presentation. Then they introduced CNN to classify the malware. In their work they shows 98.2% accuracy. They did not use any feature extraction algorithm and their approach takes very short time to detect malware. They also indicate the limitation of their that is their approach can detect the code semantics.

Hardy et al. in their work [62] a deep learning framework for malware classification. They named it DL4MD, which is based on the API call sequence of malicious programs. They transform the API calls into 32-bit global ID. For API call to ID transformation they use a API call database. The features are then feed to encoders based on deep learning platform.

Huang et al. [63] in their research combined the API calls with the function call for malware classification. He also combine tri-gram with API calls. Their work do both the detection and classification of malware family.

Davis and Wolff in their approach [64] dissemble the malware code using deep learning CNN. Then different features were extracted and used for classifiers. They collect the code imports, processors instruction to make fixed length of feature vector.

Tobiyama in their work [65] consider more information than API call sequence, they used the process information like ID, naming, directory etc. The collected information is then feed into RNN to create feature image. The collected images are then feed into CNN to classify the malware.

All the above proposal for deep learning approach have higher accuracy than the classical machine learning algorithm but deep learning comes with its own drawbacks, that is these algorithm need more time to train models and huge amount of data is required for better accuracy.

17 Malware Detection in the Android Mobile

We are separately focusing on malware programs on android mobiles is because the use of android mobiles is creating a change in the information security. Android OS is becoming very popular as it can run very powerful applications comparatively to other mobile OS. Again, Moreover, due to its openness and user friendly nature it is even gaining more popularity. But, due to the growing popularity it is being used by the cyber attackers to carry out illegitimate operations. Due to openness nature malicious codes can be easily integrated with the android applications which has increased the security concern at a high level.

Many researcher have proposed different approaches to malware detection in android platform. We will not mention some of the latest research proposals only. The main purpose of reviewing these separately is the scope of future research.

Some of the proposal based on classical analysis of android malware can be found in [66–75], where different statical and dynamic analysis is discussed in details. These signature based algorithms have very low false positive rate for known signatures but they fails in case of unknown signatures. In case of anomaly based detection, they can detect unseen malware but highly false positive.

Some the deep learning approach for detecting malicious code in android is also proposed by some of the researchers over the past years. Authors of [65] proposed a deep learning based malware detection for android platform using deep neural network consisting of two stage. The result shows high accuracy but the experiment was only carried out in smaller datasets.

Authors of [76] also presented a method based on hidden markov model and structural entropy. Their work performs well on detecting unknown malware but accuracy could be increased using larger dataset.

Authors of [77] also presented deep learning based approach based on sequence classification. Their algorithm learns from the API call sequences to classify between benign and malevolent apps. Main drawback of their work is that this approach can not detect malware when is downloaded and executed at runtime.

In [78] authors proposed a framework for malevolent code on android platform. The authors also train their model using API calls. Their work shows better accuracy than their previous work presented at [77] but the main drawback is that this framework does not consider native codes.

In [79] authors presented deep learning approach based on malware characteristics extracted by association rule mining techniques. In their work they showed that fine-grained feature leads to higher accuracy in malware detection.

Although, some significant proposals based on classical machine learning and signature based detection has been presented but there is lot of scope for introducing deep learning algorithms. Most the deep learning based proposals were not so accurate for native codes and dynamic code detection, which opens a lot of scope for future research.

18 Future Direction and Conclusion

In this study of malware and malware detection methods, our main focus was to highlight the future research scopes of integrating deep learning and malware detection. As, malware programs are evolving at a high speed and also due to the availability of technology creation of malware programs are getting easy too. With advance obfuscation techniques malware programs are outperforming the traditional anti malware detectors. Even though deep learning algorithms poses better accuracy over classical machine learning algorithms, less research has been carried out in fields of malware detection in android platform. Moreover, The main challenge in introducing deep learning to classify malware is that these algorithms need large data set and time to learn which need to be addressed. We presented this survey in a manner which will be a key reference to future research.

References

1. McGraw G, Morrisett G (2000) Attacking malicious code: a report to the infosec research council. IEEE Softw 17(5):33–41
2. Xufang L, Loh PKK, Tan F (2011) Mechanisms of polymorphic and metamorphic viruses. In 2011 European intelligence and security informatics conference (EISIC) 149–154
3. Cohen F (1987) Computer viruses. Comput Secur 6:22–35
4. EroCarrera, Silberman P (2010) State of malware: family ties
5. Egele M et al (2008) A survey on automated dynamic malware-analysis techniques and tools. ACM Comput Surv 44(2):1–42
6. Vinod P et al (2009) Survey on malware detection methods
7. WebSource: https://www.cisco.com/c/en/us/about/security-center/virus-differences.html
8. Yin H et al (2007) Panorama: capturing system-wide information flow for malware detection and analysis. In: Proceedings of the 14th ACM conference on computer and communications security. ACM, Alexandria, pp 116–127
9. Idika N, Mathur AP (2007) A survey of malware detection techniques
10. Bostami B, Ahmed M, Choudhury S (2019) False data injection attacks in internet of things. In: Al-Turjman F (ed) Performability in internet of things. EAI/Springer innovations in communication and computing. Springer, Cham
11. Beaucamps P (2007) Advanced polymorphic techniques. Int J Comput Sci 2(3):194–205
12. Szor P (2005) The art of computer virus research and defense. Addison-Wesley Professional, Upper Saddle River
13. Shah A (2010) Approximate disassembly using dynamic programming [PhD. Thesis], San Jose State University, US
14. Szor P (1998) The Marburg situation. Virus Bull:8–10
15. Filiol E (2005) Computer viruses: from theory to applications. Springer, Paris
16. Walenstein A, Mathur R, Chouchane M et al (2007) The design space of metamorphic malware. In: Proceedings of the 2nd international conference on information warfare and security (ICIW 2007), pp 241–248
17. Lakhotia A, Kapoor A, Kumar E (2004) Are metamorphic viruses really invincible? Virus Bull:5–7
18. Ferrie P, Corporation S, Monica S (2001) Hunting for metamorphic. Proceedings of the Virus Bulletin Conference 2001, Czech Republic, Prague, 2001 September 27–28, 123144
19. Gutmann P (2007) The commercial malware industry
20. Islam MDR, Tian R, Batten LM, Versteeg S (2013) Classification of malware based on integrated static and dynamic features. J Netw Comput Appl 36(2):646–656
21. Tahir R (2018) A study on malware and malware detection techniques. Int J Educ Manag Eng 8:20–30. https://doi.org/10.5815/ijeme.2018.02.03
22. Jacob G, Debar H, Filiol E (2008) Behavioral detection of malware: from a survey towards established taxonomy. J Comput Virol 4(3):251–266
23. Hofmeyr S, Forrest S, Somayaji A (1998) Intrusion detection using sequences of system calls. J Comput Secur 6:151–180
24. Sato I, Okazaki Y, Goto S (2002) An improved intrusion detection method based on process profiling. IPSJ J 43:3316–3326
25. Mohata VB (2013) Mobile malware detection techniques. Int J Comput Sci Eng Technol (IJCSET)
26. Schultz M, Eskin E, Zadok E, Stolfo S (2001) Data mining methods for detection of new malicious executables. In IEEE symposium on security and privacy, pp 38–49. IEEE Computer Society
27. Henchiri O, Japkowicz N (2006) A feature selection and evaluation scheme for computer virus detection. In: Proceedings of ICDM-2006, Hong Kong, pp 891–895
28. Ye Y, Wang D, Li T, Ye D (2007) IMDS: intelligent malware detection system. In: Proceedings of the ACM international conference on knowledge discovery data mining, pp 1043–1047

29. Ye Y, Li T, Jiang Q, Wang Y (2010) CIMDS: adapting post processing techniques of associative classification for malware detection. IEEE Trans Syst Man Cybern C 40(3):298–307
30. Jeong K, Lee H (2008) Code graph for malware detection. In information networking. In: ICOIN. International conference on, Jan 2008
31. Lee J, Jeong K, Lee H (2010) Detecting metamorphic malwares using computing, ser. ACM, New York, pp 1970–1977
32. Ye Y, Li T, Huang K, Jiang Q, Chen Y (2010) Hierarchical associative classifier (HAC) for malware detection from the large and imbalanced gray list. J Intell Inf Syst 35(1):1–20
33. Ahmed F, Hameed H, Shafiq MZ, Farooq M (2009) Using spatio-temporal information in API calls with machine learning algorithms for malware detection. In: AISec '09 Proceedings of the 2nd ACM workshop on Security and artificial intelligence, pp 55–62
34. Bilar D (2007) OpCodes as predictor for malware. Int J Electron Secur Digit Forensics 1(2):156
35. Santos I, Brezo F, Nieves J, Penya Y (2010) Idea: OpCode-sequencebased malware detection. In: Engineering secure software and system. Springer, Berlin/Heidelberg
36. Santos I, Brezo F, Ugarte-Pedrero X, Bringas PG (2011) OpCode sequences as representation of executables for data-mining-based unknown malware detection. Inf Sci 231:64–82
37. Santos I, Brezo F, Sanz B, Laorden C, Bringas PG (2011) Using opCode sequences in single-class learning to detect unknown malware. IET Inf Secur 5(4):220
38. Santos I, Laorden C, Bringas P (2011) Collective classification for unknown malware detection. In: Proceedings of the 6th ACM symposium on information, computer and communications security
39. Santos I, Sanz B, Laorden C (2011) OpCode-sequence-based semisupervised unknown malware detection. In: Computational intelligence in security for information systems. Springer, Berlin/Heidelberg
40. Runwal N, Low RM, Stamp M (2012) OpCode graph similarity and metamorphic detection. J Comput Virol 8(1–2):37–52
41. Shabtai A, Moskovitch R, Feher C, Dolev S, Elovici Y (2012) Detecting unknown malicious code by applying classification techniques on OpCode patterns. Secur Inf 1(1):1
42. Gerald GBS, Tesauro J, Kephart JO (1996) Neural network for computer virus recognition. IEEE Expert
43. Arnold W, Tesauro G (2000) Automatically generated Win32 heuristic virus detection. In Virus Bulletin Conference
44. Abou-assaleh, T, Cercone N, Keß V, Sweidan R (2004) N-gram-based detection of new malicious code, no. 1
45. Maloof MA, Kolter JZ (2006) Learning to detect malicious executables in the wild. In roc of the 10th ACM SIGKDD International Conference on Knowledge Discovery and Data Mining
46. Moskovitch EY, Stopel D, Feher C, Nissim N, Japkowicz N (2009) Unknown malcode detection and the imbalance problem. J Comput Virol 5(4):295–308
47. Bruschi D, Martignoni L, Monga M (2006) Detecting self-mutating malware using control-flow graph matching. In: Büschkes R, Laskov P (eds) Detection of intrusions and malware & vulnerability assessment, volume 4064 of LNCS. Springer, Berlin, pp 129–143
48. Zhao Z (2011) A virus detection scheme based on features of control flow graph. 2nd International Conference on Artificial Intelligence, Management Science and Electronic Commerce (AIMSEC), pp 943–947
49. Bonfante G, Kaczmarek M, Marion JY (2007) Control flow graphs as malware signatures. WTCV
50. Eskandari M, Hashemi S (2011) Metamorphic malware detection using control flow graph mining. Int J Comput Sci Netw Secur 11:1–6
51. Kim K, Moon BR (2010) Malware detection based on dependency graph using hybrid genetic algorithm. In Proceedings of the 12th annual conference on Genetic and evolutionary computation, July 07–11, 2010

52. Nataraj L, Karthikeyan S, Jacob G, Manjunath BS (2011) Malware images: visualization and automatic classification. In: Proceedings of the 8th international symposium on visualization for cyber security, VizSec '11. ACM.. ISBN 978-1-4503-0679-9, New York, pp 4:1–4:7. https://doi.org/10.1145/2016904.2016908
53. Dahl GE, Stokes JW, Deng L, Yu D (2013) Large-scale malware classification using random projections and neural networks. In Acoustics, Speech and Signal Processing (ICASSP), 2013 IEEE International Conference on Acoustics. IEEE, 3422–3426
54. Saxe J, Berlin K (2015) Deep neural network based malware detection using two dimensional binary program features. In 2015 10th International Conference on Malicious and Unwanted Software (MALWARE). IEEE
55. Pascanu R, Stokes JW, Sanossian H, Marinescu M, Thomas A (2015) Malware classification with recurrent networks. In acoustics, speech and signal processing (ICASSP), 2015 IEEE International Conference on Acoustics. IEEE, 1916–1920
56. Cakir B, Dogdu E (2018) Malware classification using deep learning methods. In: Proceedings of the ACMSE 2018 conference (ACMSE '18). ACM, New York. Article 10, 5 pages
57. Raff E, Barker J, Sylvester J, Brandon R, Catanzaro B, Nicholas C (2017) Malware detection by eating a whole exe. arXiv preprint arXiv:1710.09435
58. David OE, Netanyahu NS (2015) DeepSign: deep learning for automatic malware signature generation and classification. 2015 International Joint Conference on Neural Networks (IJCNN), Killarney, 2015, pp 1–8
59. Karbab E Debbabi M Derhab A Mouheb D (2017) Android malware detection using deep learning on API method sequences
60. Choi S, Jang S, Kim Y, Kim J (2017) Malware detection using malware image and deep learning. 2017 International conference on information and communication technology convergence (ICTC), Jeju, 2017, pp 1193–1195
61. Le Q, Boydell O, Mac Namee B, Scanlon M (2018) Deep learning at the shallow end: malware classification for non-domain experts. Digit Investig 26:S118–S126
62. Hardy W, Chen L, Hou S, Ye Y, Li X (2016) Dl4md: a deep learning framework for intelligent malware detection. *Athens: The Steering Committee of The World Congress in computer science, computer engineering and applied computing (WorldComp),* pp 61–67
63. Huang W, Stokes JW (2016) MtNet: a multi-task neural network for dynamic malware classification. In: In Proc. of the 13th international conference on detection of intrusions and malware, and vulnerability assessment, DIMVA 2016. Springer, Cham, pp 399–418
64. Davis A, Wolff M (2015) Deep learning on disassembly data. URL: https://www.blackhat.com/docs/us-15/materials/us-15-Davis-Deep-Learning-On-Disassembly.pdf
65. Tobiyama S, Yamaguchi Y, Shimada H, Ikuse T, Yagi T (2016) Malware detection with deep neural network using process behavior. In: 2016 IEEE 40th annual computer software and applications conference (COMPSAC), vol 2, pp 577–582. https://doi.org/10.1109/COMPSAC.2016.151
66. Kang H, Jang JW, Mohaisen A, Kim HK (2015) Detecting and classifying android malware using static analysis along with creator information. Int J Distrib Sens Netw 11(6):479174
67. Faruki P, Laxmi V, Bharmal A, Gaur MS, Ganmoor V (2015) AndroSimilar: robust signature for detecting variants of Android malware. J Inf Secur Appl 22:66–80
68. Song J, Han C, Wang K, Zhao J, Ranjan R, Wang L (2016) An integrated static detection and analysis framework for Android. Pervasive Mob Comput 32:15–25
69. Sun M, Li X, Lui JC, Ma RT, Liang Z (2017) Monet: a user-oriented behavior-based malware variants detection system for Android. IEEE Trans Inf Forensics Secur 12(5):1103–1112
70. Rovelli P, Vigfússon Ý (2014) PMDS: permission-based malware detection system. In: Prakash A, Shyamasundar R (eds) ICISS 2014. LNCS, vol 8880. Springer, Cham, pp 338–357. https://doi.org/10.1007/978-3-319-13841-1_19
71. Wu DJ, Mao CH, Wei TE, Lee HM, Wu KP (2012) DroidMat: android malware detection through manifest and API calls tracing. In: 2012 seventh Asia joint conference on information security (Asia JCIS), pp. 62–69. IEEE

72. Talha KA, Alper DI, Aydin C (2015) APK auditor: permission-based Android malware detection system. Digit Investig 13:1–14
73. Sato R, Chiba D, Goto S (2013) Detecting Android malware by analyzing manifest files. Proc Asia Pac Adv Netw 36(23–31):17
74. Ping X, Xiaofeng W, Wenjia N, Tianqing Z, Gang L (2014) Android malware detection with contrasting permission patterns. China Commun 11(8):1–14
75. Vidal JM, Monge MAS, Villalba LJG (2018) A novel pattern recognition system for detecting Android malware by analyzing suspicious boot sequences. Knowl-Based Syst 150: 198–217
76. Canfora G, Mercaldo F, Visaggio CA (2016) An HMM and structural entropy based detector for android malware: an empirical study. Comput Secur 61:1–18
77. Karbab EB et al (2017) Android malware detection using deep learning on API method sequences. *CoRR* abs/1712.08996: n. Pag
78. Karbab E, Debbabi M, Derhab A, Mouheb D (2018) MalDozer: automatic framework for android malware detection using deep learning. Digit Investig 24:S48–S59. https://doi.org/10.1016/j.diin.2018.01.007
79. Yuan Z, Lu Y, Xue Y (2016) Droiddetector: android malware characterization and detection using deep learning. Tsinghua Sci Technol 21(1):114–123

Biozid Bostami attained his Bachelor of Science Degree in Computer Science and Information Technology with High Distinction from Islamic University of Technology, OIC. He is working in the area of Big Data Mining, Machine Learning and Network Security in collaboration with researchers from Australian and Canadian universities. He is working to develop efficient and accurate Anomaly Detection techniques for network traffic analysis to handle the emerging Big Data problems.

Mohiuddin Ahmed attained his PhD from UNSW Australia and currently working as Lecturer of Computing and Security within School of Science at Edith Cowan University. His research interests include Big Data Mining, Machine Learning and Network Security. He is working to develop efficient and accurate Anomaly Detection techniques for network traffic analysis to handle the emerging Big Data problems. He has made practical and theoretical contribution for data summarization for network traffic analysis. His research also has a high impact on critical infrastructure protection (SCADA systems, Smart Grid), information security against DoS attacks and complicated health data (heart disease, nutrition) analysis. He has published a number of journals and conferences papers in reputed venues of computer science. Mohiuddin holds a Bachelor of Science Degree in Computer Science and Information Technology with High Distinction from Islamic University of Technology, OIC.

The Utilization of Blockchain for Enhancing Big Data Security and Veracity

Satriyo Wibowo and Arwin Datumaya Wahyudi Sumari

Abstract Blockchain as one of technological hype in digital economy besides the Internet-of-Things (IoT) and Big Data Analytic, fills the need of a secured peer-to-peer connection with the concept of distributed database. However, it does not eliminate the centralized database on massive data storage which it is the core of Big Data. Blockchain is more suitable for information log, a kind of application that requires dynamic and updated information with hierarchical hash security features to support a distributed database system. The Blockchain features are prospective to enhance the security of Big Data from attacks to its CIA Triad, namely Confidentiality, Integrity, and Availability. As information has become a crucial and critical to business, meanwhile managing a huge-volume data is also challenging in terms of its security and veracity, therefore Blockchain technology can be considered as a prospective solution. Based on our study, we found that Blockchain can enhance the security of Big Data by strengthening the security of the data storage, enhancing the data integrity using digital certificate and chaining the block using hash of previous block, and enhancing data availability using peer-to-peer transmission, distributed nodes, and consensus method. Blockchain can also enhance the performance of Big Data Analytic by providing a better data veracity from token-based validation to enhance the truth discovery, and ID decentralization to prove the identity of data source.

Keywords Availability · Big data · Blockchain · Confidentiality · Identity decentralization · Information security · Integrity · Veracity · Provenance · Token-based validation

---

S. Wibowo (✉)
Indonesia Cyber Security Forum (ICSF), Pesanggrahan, South Jakarta, Indonesia
e-mail: satriyowibowo@icsf.or.id

A. D. W. Sumari
Department of Electrical Engineering, State Polytechnic of Malang, Malang, East Java, Indonesia

Faculty of Defense Technology, Indonesia Peace and Security Center (IPSC), Indonesia Defense University, Sentul, East Java, Indonesia
e-mail: arwin.sumari@polinema.ac.id

Z. Md. Fadlullah, A.-S. Khan Pathan (eds.), *Combating Security Challenges in the Age of Big Data*, Advanced Sciences and Technologies for Security Applications, https://doi.org/10.1007/978-3-030-35642-2_8

1 Introduction

Abundant data from mobile platforms, social media, and the Internet of Things (IoT) has brought complexity to the data. Since its introduction, social media has become the most used media for sending and broadcasting any form of data (text, image, video, and voice) with the help of mobile platforms as well as non-mobile ones such as laptop and Personal Computers (PCs). With the huge number of social media users all over the world, those activities have created a huge volume of data in cyberspace. Meanwhile, the increase used of IoT also contributes in the increasing volume of data in such space. This huge-volume data is not just structured data, but also unstructured one. It is also understood that within such data, there may be valuable information which is hidden in it and this information may bring fortune to business. The data such this is called as Big Data, which needs special treatment compared to traditional database. Big Data analytics is the use of advanced analytic techniques against very large, diverse data sets that include structured, semi-structured and unstructured data, from different sources, and in different sizes from terabytes to zettabytes [1].

Big Data is a data set whose size or type is beyond the ability of traditional relational database and it needs a specific method or technique to capture, manage, and process it. Big Data has one or more of the following characteristics, namely high volume, high velocity, high variety, or veracity. Volume and velocity characteritics state how much and how soon the data is generated, then variety characteristic states the condition of the data, structured or not, and veracity characteristic speaks about the level of trust in data validity. Big data comes from many types of sensors, devices, video/audio, networks, log files, transactional applications, web, and social media, which is usually generated in real time and in a very large scale.

Analysts, researchers, and business users can make better and faster decisions using the results of the analysis of the data that was previously inaccessible or thought as unusable. There are many advanced analytics techniques to be used such as text analytics, Artificial Intelligence (AI) such as machine learning and Artificial Neural Networks (ANN), predictive analytics, data mining, statistics, and natural language processing, to extract new information or gain new insights from older abandoned data or previously untapped data sources resulting in better and faster decisions.

Data is the new oil. Institution that owns data and is capable of making use it will gain more benefits than others. In the age of information, protecting the security of the data and having trusted level of it, has become the most essential part in business. Therefore, the technology to enhance the security and the veracity of the data is critical to ensure the continuity of business.

2 Big Data Security

With Big Data comes bigger responsibilities. The most discussed issue in Big Data is the security of the information and the protection of the privacy in its platforms. Since all the tools used in the industry are relatively new and there is no recent Structured Query Language (SQL) database too, it brings a high risk for potential attacks and security breaches since none of those tools equipped with robust built-in security policies and mechanisms.

Generally, the systems are not compromised due to the exploitation of the communication protocols vulnerabilities or the cryptographic primitives. Most of the breaches come from bad configuration of the access controls and the authentication policies. So the authentication and the access control technologies are the main elements to address the security and the privacy issues in Big Data. Actually, any effective access control system should satisfy the main security properties namely, Confidentiality, Integrity, and Availability or abbreviated as CIA and called as CIA Triad.

Integrating Big data and Cloud computing has been gaining attention recent days because of its cost effective and operational easiness even though sacrificing control. The fundamental issue that should be considered is the security of the Big Data cloud environment. There are some security vulnerabilities that rise in their integration, and it is creating a new unfamiliar platform. One of the most known Big Data cloud security vulnerability is platform heterogeneity. There are many Big Data deployments that require deploying a new platform in the cloud while the existing security tools and practices will not work for such platform because new security tools are needed to be developed. These security tools could include authentication, access control, encryption, intrusion detection, and event logging and monitoring. In addition to the security policies, the Big Data consolidation plans should be taken into consideration while the integration with the cloud environment [2].

It seems like every week business, a new company must notify its customers that their data may have been compromised, or a personal information may have been affected. Data breaches can happen for various reasons such as data can be mishandled or sold to the third parties, or holes in a website's security system can leave information unprotected. One of the latest victims was Marriott hotel, which recently revealed that hackers had been able to access the information of an estimated 500 million customers. Some of the biggest victims in 2018 include T-Mobile, Quora, Google, and Orbitz. Facebook dealt with a slew of major breaches and incidents that affected more than 100 million users of the popular social network [3]. Huge-volume data needs security more than just the traditional one. In this paper, we are proposing an approach to enhance Big Data security by apprehending that the security threats are evolving and more dangerous than before.

3 Big Data Veracity

Data is generated at a tremendous pace and there must be enough measures in place to verify the nature of Big Data. A research has shown that 80% of data in the Big Data is uncertain one [4]. An analysis carried out on such uncertain data may lead to untrusted result and shape poor decisions. The aspect of Big Data that deals with its correctness is known as Big Data veracity. Veracity means that the data can be trusted. However, the data can also be comprehended only to some extent or to some confidence level, such the data observed by sensor that is only comprehended up to some level.

According to IBM as depicted in Fig. 1, the red curve shows the proportion of the data whose veracity is unknown. By the end of 2015 the percentage of uncertain data approximately reached 80%. The uncertainty of the veracity of the data can come from various sources. For instance, the measurement error of sensors or lack of credential of social media. From this point of view, if there is no solution, it is becoming harder and harder to trust the analysis of Big Data because the low of data veracity. In this paper, we are also proposing an approach to enhance the Big Data veracity by apprehending that its veracity is a warranty for a trusted result of its analysis.

3.1 Problem Statement

The rise of Blockchain technology is encouraged by Bitcoin fame in 2017. Even though its value now is already stable below USD 4000, Bitcoin ever reached USD 19,000 at the end of 2017. This causes many institution started to learn other bene-

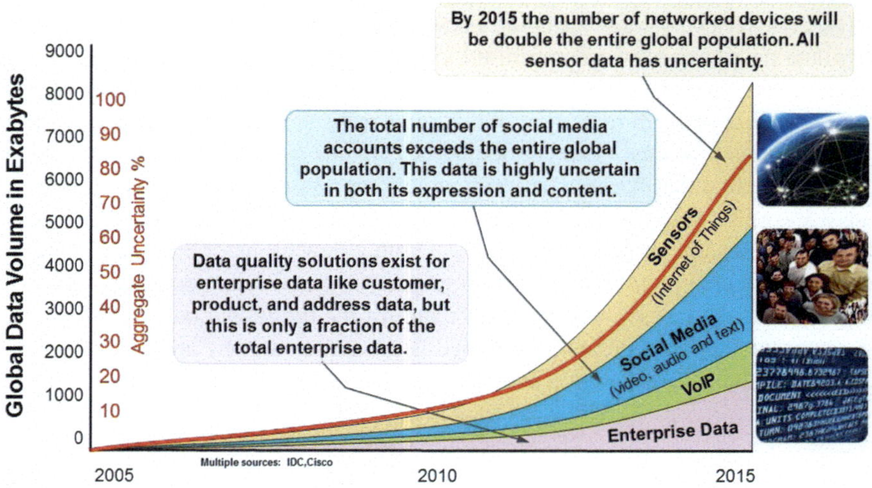

Fig. 1 Data veracity [4]

fits of Blockchain implementation. Can Blockchain give a solution for enhancing Big Data security and veracity? That question is the problem statement of this paper. In the effort to deliver solutions to the problem, we will dig in detail what Blockchain can do for Big Data security and veracity in the next following sections.

4 Blockchain

Three technologies hype that supports 4.0 industry jargon and digital economy are Big Data Analytic, IoT, and Blockchain. IoT takes the responsibility on the data capture and Blockchain acts as the data storage based on secure peer-to-peer connections of distributed databases. However, it will not eliminate the centralized database, especially on massive data storage that fuels Big Data. Blockchain is more suitable for applications that require logs of information that always moves and updates compared to quality passive data or massive data on traditional database.

As defined by the World Economic Forum (WEF), "Blockchain technology allows parties to transfer assets to each other in a secure way without intermediaries. It enables transparency, immutable records, and autonomous execution of business rules." Blockchain actually is a part of database technology, however have a distinctive working concept like: distributed database, peer-to-peer transmission, transparency, irreversible notes, and computational logic. Using a decentralized concept, Blockchain uses the concept of consensus to decide a transaction. Because of its nature, this solution is not for every problem. Blockchain Implementation Assessment Framework (BIAF) is used for assessing suitable problem for Blockchain solution and how to implement it.

4.1 Blockchain Based on Database Technology

Blockchain is a distributed-type database technology. Distributed database is a type of database where the data is stored across multiple computing devices. Moreover, Blockchain is a distributed ledger, a type of distributed database that assumes the possible presence of malicious users (nodes) and composed of a chain of cryptographically linked 'blocks' containing batched transactions and generally broadcasts all data to all participants in the network. It has a data structure like paper with page numbers arranged in a book. When a new page is added and filled, the previous pages' contents cannot be changed anymore [5].

Based on the read and write access, Blockchain generally are divided into many platforms. 'Read' access refers to who can access a distributed ledger network and sees the transactions. It is called *Public* if anybody can access the ledger and sees the transactions, or *Private* if only selected parties are able to access the ledger and see the transactions. 'Write/Commit' access refers to who can take part in making changes to a distributed ledger (e.g., who can add blocks to a Blockchain). It is

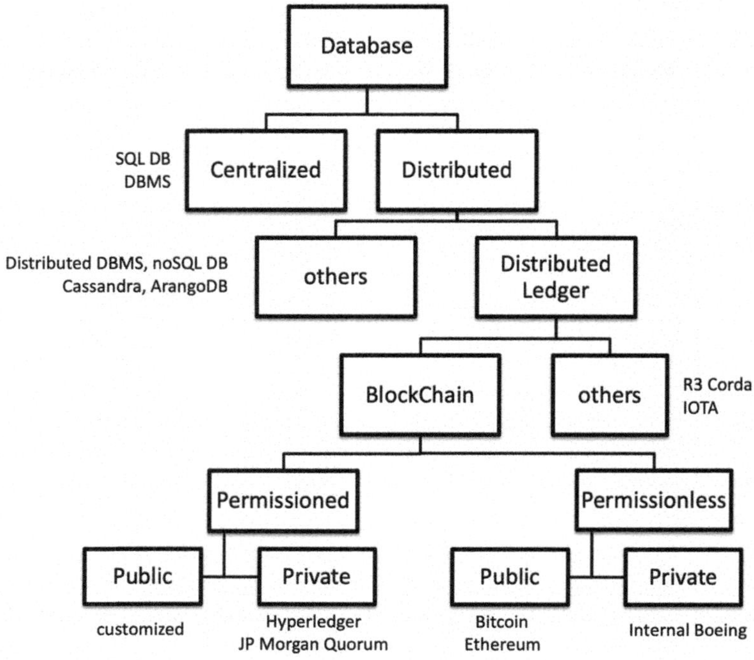

Fig. 2 Database Tree and the position of Blockchain in this tree [6]

called *Permissionless (open)* that is, anyone can, in theory, participate in the consensus process (in practice, however, often limited by resource requirements such as owning suitable hardware or cryptocurrency) or *Permissioned (closed)* that is, only selected parties can make changes to the distributed ledger. These things greatly influence the form of consensus, data confidentiality, and transaction speed so that the selection of a platform is crucial.

The following chart (Fig. 2) is the Blockchain in the database technology tree compiled by Blockchain Nusantara Research Group from several sources [6]. Blockchain Nusantara Research Group or Blockchain Nusantara is a research group established in Jakarta, Indonesia and Melbourne, Australia focusing on Blockchain research where the first author is one of its founder. "Nusantara" means "the Indonesian Archipelago".

From the tree, we can see that Blockchain is located on Distributed Ledger branch and it has two more branches where each branch has other two branches. They are as follows:

(a) Public Permissioned. It this sheme, the data is transparent and anyone can read it. However, not anyone can join the network and write the data. It is a special case when a consortium needs to control public information.
(b) Private Permissioned. In this scheme, the data is restricted to read and write. Only parties with permission that have the access to do that especially if it is a private, corporate, or state classified information. Hyperledger and Quorum

platforms are developed based on this method. The first one is built on open source platform than the latter one.

(c) Public Permissionless. The data is transparent where anyone can read and write it even though the privacy is secured by pseudoanonym concept.

(d) Private Permissionless. In this scheme, the permission to join the network is more flexible but not anyone can read the data. Boeing develops Blockchain with this concept for its internal use.

The Private Permissioned (Private Blockchain) and Public Permissionless (Public Blockchain) are platforms that have more attention. Those two concepts will be discussion in this paper along with example of use-case to give more understanding to the readers.

4.2 The Concept of Blockchain

In general, Blockchain works like a distributed database, carries out peer-to-peer transmission, transparent, irreversible notes, and computational logic. Distributed database means that each party that joins Blockchain has access to all data and completed transaction history without exception. It is a form of pure transparency and adopts a decentralized database system. Each party may verify its partner transaction directly without a middleman.

Peer-to-peer transmission means that the communication or the transaction occurs between one party and another is done without intermediary node. Each node can store and forward information to another node. Transparency with pseudo-anonymity (idle identity (ID), especially on Public Blockchain) means that such node or user in Blockchain has an address that contains 30 alphanumeric characters or more for user identification marks (such as username id). Users can choose to keep their original name hidden or visible when making a transaction. Irreversible notes or finality means that if the transaction has been recorded in the database, then the record cannot be changed because of Blockchain's security system. The system comes with a variety of algorithms that make all transactions in chronological order and accessible to all parties. If you want to change the data of one transaction, then all data must be deleted and you have to start all from the scratch.

Computational logic means that Blockchain can be programmed specifically so that transactions can be automatically performed when a criterion has been met. For example, companies can program their Blockchain accounts for automatic payment of a procurement of raw material when the material carrier truck has entered the company's Headquarters (HQ). Ccomputational logic is also called as smart contract but not all platform has this capability. Bitcoin as a platform only focuses on transaction capability. Ethereum is a public platform that introduces the concept of smart contract and many Initial Coin Offering (ICO) is built based on it. If transaction is just the act of sell and buy, so in the smart contract there will be *if… then…* to make the transaction occurred.

4.3 The Consensus in Blockchain

Just like the crowd, there must be a consensus on how to make a decision on something. Because of the different in the nature, the consensus on Public Blockchain is very different from the Private Blockchain. Public Blockchain exists on untrusted network where anybody can join the network and can read and write the ledger. Because of that the consensus is rather difficult to filter the fraudster and hacker. There are three base models of consensus known and used in Public Blockchain, namely Proof of Work (PoW), Proof of Stake (PoS), and Proof of Assignment (PoA) [7].

PoW is used by cryptocurrency pioneer like Bitcoin, Ethereum, and Litecoin. They have miner concept, the party that creates block for containing the transaction. To choose which miner that has accepted blocks, the system creates mathematical problem to solve. Whoever solves that equation, their blocks are accepted and got the payment. This method is proven resilience against the internal and external attacks, however it needs high energy consumption. Moreover, only the miner with huge processing power wins the race and it creates centralize power among the system.

PoS tries to overcome the problem of energy consumption by using the stake concept so that the party that provides higher stakes will win and authorizes the transaction. The system actively penalizes dishonest behavior among validators, but it will increase risk of forks. Forks happen if many parties of miners already create chained blocks so long that unwillingly give up of consensus decision. Hard fork creates a new coin beside the original one. This method is used by Redcoin and NavCoin. PoA uses another method by assigning a party to be validator. This method speeds up the transaction, but the system will be a centralized and is prone to be attacked. This method is used by IOTW.

The consensus on Private Blockchain is much easier because it is naturally built on a trusted network. However, it still uses a Byzantine Fault Tolerant (BFT) to make sure all the information is delivered and there is no compromised node that can influence the validation. Hyperledger as private platform uses different model of consensus on each platform. Fabric uses Practical BFT (PBFT), Indy uses Redundant BFT (RBFT), Iroha uses Sumeragi, and Sawtooth uses Proof of Elapsed Time (PoET). Their consensus are based on crash tolerant, election based, reputation based, or lottery based. Naturally, this kind of consensus is faster that Public Blockchain [8].

4.4 Blockchain Implementation Assessment Framework

As stated above, Blockchain is a fresh addition for a complex system of data storage that provide a solution for certain problems. Data stored at the distributed ledger is considered more secure than the centralized one that is prone to have vulnerability

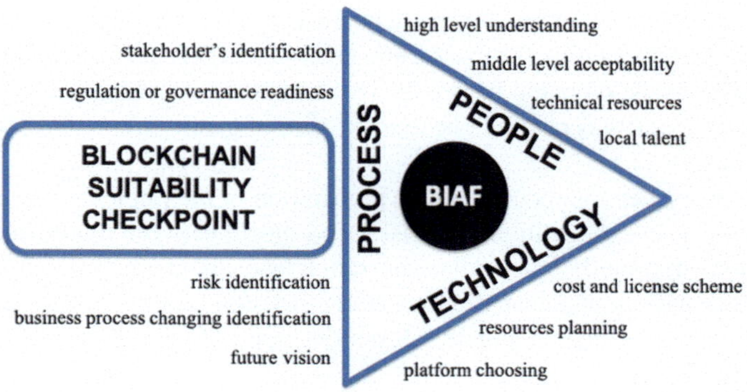

Fig. 3 Blockchain Implementation Assessment Framework [9]

because of its single point of failure. Certainly, not all data can be treated like that, there will be an assessment which problem can be solved by Blockchain.

BIAF, as depicted in Fig. 3, is a framework for measuring suitability of a use-case with Blockchain implementation [9]. It is built for a general purpose because not all problem can be solved by Blockchain. For example, even though logistic, supply chain management, and remittance are ideal use-cases, but if the regulation does not take a place yet, then the Blockchain implementation surely will suffer. There are two level of assessment. The first one is a set of question to make sure the case is a suitable case for Blockchain implementation. The second one is a framework built from the triangle of Process-People-Technology framework [10]. For each kind, there will be set of assessment to ensure that everything is ready before implementing the Blockchain. These are the checkpoints to ascertain how Blockchain could be the solution to problems, as follows.

1. centralized database solution is considered a deficiency, maybe because it is less secure, too many synchronization process, hard to track data history, and so on;
2. the asset, or key data that moves across the system and will be changed by transaction, is a historical asset that its history matters;
3. the system needs single database with redundant dataset so that all stakeholder can improve the performance altogether;
4. there are multi-stakeholders that hold different sets of data but complement to each other to enrich the asset;
5. the location of each data source is spread over many locations;
6. a lot of participants who need to change the data, but only certain participants can change basic application;
7. there is personal data protection requirement.

The most important checkpoints are the asset and multi-stakeholder because Blockchain will be seen as a giant database by all nodes, even though there are many stakeholders that have their own datasets. Moreover, Blockchain can be set to permit the communication on specific group not to be accessed by others. The second level is performed after there are conclusions that the case is a suitable for Blockchain implementation. The assessment is based on Process-People-Technology method that must be prepared before the implementation could take place. It is not all serial steps, mostly is parallel one to make sure of gaining enough momentum to move forward.

Process assessment is focused on the preparation of the implementation process to identify the problems, mitigate the risks, coordinate between multi stakeholders, and the most importantly buy in all organization level about Blockchain. There will be a lot of discussions and changing procedure that affect the organization behavior, structure, process business, and tasks. From this point of view, there will be checkpoints to fill such as:

1. stakeholder's identification, to identify multi-stakeholder that exist and have interest on the data;
2. regulation or governance readiness to ensure that all of stakeholder is already to bind with the regulation or the contract to store the data;
3. risk identification of the implementation that will be occured;
4. identification of the business process that can be changed after the implementation;
5. the future vision of the system so that there will be no mistake in choosing the right platform.

People assessment is focused on the preparation of human resource, which is the most important part of an organization. Everybody on the board must understand the risks, cost, and benefit of the Blockchain implementation. It is important that the organization has enough resources and understanding on it as well. Moreover, it is crucial that the internal organization resources have capabilities on Blockchain in order to avoid any dependency on the external ones. There will be checkpoints to fill such as:

1. high level understanding, a very important session to make sure there will be no hindrance from the leader, and the cost and benefit of this implementation is already accepted;
2. middle level acceptability is important because they are the person to implement and maintain the system;
3. technical resources, to make sure there will be an assistance when needed;
4. local talent, to increase the independency from the outsiders.

Technology assessment is focused on the technological solution for the use-case. Blockchain is still a growing technology and looking for its best suitability. Choosing a platform that has future-proof, has firm support from community and business, and has capabilities to grow is very important. So that, there will be checkpoints to fill such as:

1. platform choosing, because many platform are competing and not all resources understand the other;
2. resources planning, maintaining human resources, the cost and time to develop;

3. cost and license scheme.

This kind of assessment needs careful and lengthy discussion with the problem owner and the most time costing is regulation or governance readiness. For the private sector, this assessment may not be too difficult as long as there is a business between stakeholders. However, for public sector, this part seems forever because they must comply the regulations. If there is no regulation or worse, the regulation is against this move, then this will be the highest hindrance for the Blockchain implementation.

5 The Blockchain's Role in Big Data Security

As stated above, the authentication and access control technologies are the main elements to address the security and privacy issues in Big Data. Es-Samaali, Outchakoutch, and Leroy [11] proposed a Blockchain-based access control for Big Data and demonstrated the feasibility of using blockchain technology to manage access control process for Big Data through the description of their proposed framework. It leverages the salient features of Blockchain that are, distribution, full-fledged and append-only ledger to make a promising solution for addressing the aforementioned access control challenges in Big Data. However, adopting the Blockchain technology to handle access control functions is not straightforward and additional critical issues emerge.

Although several parties claim that Blockchain is a technology that disrupts security concepts, however, large-scale robbery cases against Blockchain-based cryptos are still occurring. Disruption of the security concept at Blockchain is only on its storage concept. The data is stored in the block, and then the hash of the payload inserted to be the header on the next block. Changing the data is no longer possible and the longer the block is made, it is difficult to do data destruction. This security feature can be utilized for Big Data security. Attacks on a system usually lead to attack tthree things in information security known as CIA triad: Confidentiality, Integrity, and Availability. CIA Triad is a model to guide the policy of information security in an organization. This term is stated on ISO/IEC 27001:2013 as components to protect and secure the information. Attacks on the confidentiality of data are either forced or veiled attempts to steal data and information. Attacks on data integrity try to change the contents of the data and information for the benefit of the attacker, while attacks on availability make such data or information inaccessible to cause material loss. From the descriptions above, data integrity in Blockchain is maintained by chaining the blocks and its availability is secured by the consensus method.

Figure 4 below shows a case example on Hyperledger Sawtooth. The idea is simple, use "intkey create_batch" command to prepare batches of transactions that set a few keys to random values, which then randomly increment and decrement those values. These batches are saved locally in batches.intkey file. There are two

```
satriyowibowo@ubuntu:~$ intkey create_batch --count 10 --key-count 5
Writing to batches.intkey...
satriyowibowo@ubuntu:~$ intkey load -f batches.intkey
batches: 11 batch/sec: 196.66806197915557
satriyowibowo@ubuntu:~$ sudo bash -c "tail -10 /var/log/sawtooth/intkey-*-debug.log"
[sudo] password for satriyowibowo:
[10:10:31.335 [MainThread] core DEBUG] received message of type: TP_PROCESS_REQUEST
[10:10:31.336 [MainThread] handler DEBUG] Incrementing "ABM's" by 2
[10:10:31.339 [MainThread] core DEBUG] received message of type: TP_PROCESS_REQUEST
[10:10:31.340 [MainThread] handler DEBUG] Decrementing "A's" by 1
[10:10:31.343 [MainThread] core DEBUG] received message of type: TP_PROCESS_REQUEST
[10:10:31.345 [MainThread] handler DEBUG] Decrementing "ABM's" by 1
[10:10:31.348 [MainThread] core DEBUG] received message of type: TP_PROCESS_REQUEST
[10:10:31.349 [MainThread] handler DEBUG] Incrementing "ABM's" by 7
[10:10:31.352 [MainThread] core DEBUG] received message of type: TP_PROCESS_REQUEST
[10:10:31.353 [MainThread] handler DEBUG] Decrementing "A's" by 8
satriyowibowo@ubuntu:~$ sawtooth block list
NUM  BLOCK_ID                                                                                                                          BATS  TXNS  SIGNER
1    f46a2c9e003c760fd1a8d54b4310d9e22a5946fa495631411277fc77f4e8734258b5cdfac1fd1f603fb9695ee864416b6f1cc2410735435518310 11a3ae7be54 11    66    02158e...
0    7b5ceb65db24372b3a9ebc7a6b383bc2844cb70f6ea7a4ff3bbae95523247dc875d31ba869dd391fabd80bfa3f5da8a33f7ea1c8976b5f884d36829b605a7444  1     1     02158e...
```

Fig. 4 Blockchain Case on Hyperledger Sawtooth

blocks which are formed from the command. Figure 5 shows the information of block ID number 0. The discussion about that will be explained further more especially on confidentiality and integrity sub sections.

5.1 The Security on Confidentiality

Confidentiality is about protecting sensitive and private information from unauthorized accesses. Ayyub and Mohammad [12] are worried about confidentiality in Blockchain. They pointed on Bitcoin that has pseudo anonym method so that everybody can investigate the value inside a wallet even though the owner ID is unknown. For example, police can track wallet address and its value movement used by WannaCry ransomware but cannot detain the culprit if they keep the bitcoin and not change it to fiat money. This is the characteristic of Public Permissionless Blockchain. It is different from Private Permissioned Blockchain that is designed to be more private and secure. Therefore without permission, nobody can read and write the data, or even join the network. This kind of platform fits for corporate, business, or government use-cases even though they must build their own network.

Private Blockchain ensures only those who have permission to read and write data who can access the data. Public Blockchain is more transparent but nobody knows the owner and it is as pseudo-anonym. Attacks on one point will not cause the collapse of the system because there are still many other living and synchronized points. Blockchain increases the data protection by using encryption technology. Figure 5 shows the payload is encrypted, even though it is only a simple data containing command to increment ABM's by 2. The payload then will be hashed by SHA512 algorithm to be written on the header of the next block, and form a chain between the two blocks.

5.2 The Security on Integrity

Data Integrity is about designing a protection to the data from deletion or modification by any unauthorized party but ensuring an authorized person can reverse the damage he has done. Public Blockchain is like a big book where all transactions are transparent and can be checked by anyone to ensure the transactions' credibility. Their data structure is cannot be changed after, it can only be added. Each data from the Blockchain is connected to each other where if there is a change in one of the data blocks it will affect the next data.

From Fig. 6, it can seen that every transaction is saved on a block with a certain size so that one block can contain zero or more transactions and some additional metadata. A Blockchain is made up of a series of blocks with new blocks always added to the end. Each block has a header that contain hash information of the previous block and it is how the block achieves immutability. When a new block is

```
satriyowibowo@ubuntu:~$ sawtooth block show 7b5ceb65db24372b3a9ebc7a6b383bc2844cb70f6ea7a4ff3bbae95523247dc875d31ba869dd391fabd80bfa3f5da8a33f7ea1c8976b5f884d36829b605
a7444
batches:
- header:
    signer_public_key: 02455323db96d774fac48438f918e74bf5435a1bcdbff2e776e66e4bd8f4efbb24
    transaction_ids:
    - 0f09d14db12777b09149cb23564657f4398624f5080430f365d6d5bf636294ee036d194ba68dda565ac2370389815a043e5dc075b9f9b9372086f356c034e13b
  header_signature: 33eed4aaf75cf06645385f842f54914e369f992af98e8237b4461b196a25a039288b766f19357d66cd413fde0244f0a522eac7aa93ec9562cc27867eb78151bc
  trace: false
  transactions:
  - header:
      batcher_public_key: 02455323db96d774fac48438f918e74bf5435a1bcdbff2e776e66e4bd8f4efbb24
      dependencies: []
      family_name: sawtooth_settings
      family_version: '1.0'
      inputs:
      - 000000a87cb5eafdcca6a8cde0fb0dec1400c5ab274474a6aa82c1c0cbf0fbcaf64c0b
      - 000000a87cb5eafdcca6a8cde0fb0dec1400c5ab274474a6aa82c12840f169a04216b7
      - 000000a87cb5eafdcca6a8cde0fb0dec1400c5ab274474a6aa82c1918142591ba4e8a7
      - 000000a87cb5eafdcca6a8cde0fb0dec1400c5ab274474a6aa82c12840f169a04216b7
      nonce: ''
      outputs:
      - 000000a87cb5eafdcca6a8cde0fb0dec1400c5ab274474a6aa82c1c0cbf0fbcaf64c0b
      - 000000a87cb5eafdcca6a8cde0fb0dec1400c5ab274474a6aa82c12840f169a04216b7
      payload_sha512: 6ecd2405acff8b77d9a790f7e9cb0fe01e5974dbbb31913ab0f906f8a20406529b8326258138fb5941fb37fc670fad1cdd1e076017d91c40670236542985a7c2
      signer_public_key: 02455323db96d774fac48438f918e74bf5435a1bcdbff2e776e66e4bd8f4efbb24
    header_signature: 0f09d14db12777b09149cb23564657f4398624f5080430f365d6d5bf636294ee036d194ba68dda565ac2370389815a043e5dc075b9f9b9372086f356c034e13b
    payload: CAESgAEKJnNhd3Rvb3RoLnNldHRpbmdzLnZvdGUuYXV0aG9yaXplZF9rZXlzEkIwMjQ1NTMyM2RiOTZkNzc0ZmFjNDg0MzhmOTE4ZTc0YmY1NDM1YTFiY2RiZmYyZTc3NmU2NmU0YmQ4ZjRlZmJiMjQaEj
B4NGMxY2RmMWY5MzE4Mzc0MQ==
header:
  batch_ids:
  - 33eed4aaf75cf06645385f842f54914e369f992af98e8237b4461b196a25a039288b766f19357d66cd413fde0244f0a522eac7aa93ec9562cc27867eb78151bc
  block_num: '0'
  consensus: R2VuZXNpcw==
  previous_block_id: '0000000000000000'
  signer_public_key: 02158e99a5a95d3108765559eb9dcf193a0bad93d4f956b427d86e851c3d52cb44
  state_root_hash: 517a9071b3eb0bf18d945b80795af1f7fd0c33b2052755a7a1e709795472be42
header_signature: 7b5ceb65db24372b3a9ebc7a6b383bc2844cb70f6ea7a4ff3bbae95523247dc875d31ba869dd391fabd80bfa3f5da8a33f7ea1c8976b5f884d36829b605a7444
```


Fig. 5 Sample of Blockchain Block on Hyperledger Sawtooth

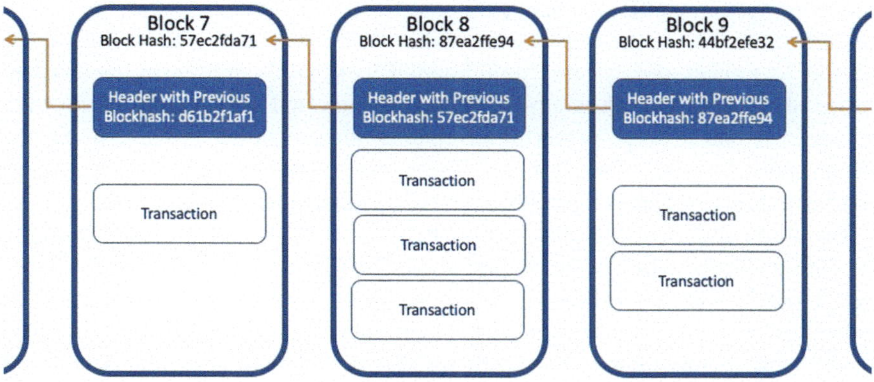

Fig. 6 Block on chain [8]

created, the process will be repeated again and again, to make a chain of blocks containing transactions. The writing of these blocks is stopped if there are mistakes on the data input which we cannot edit. We must put a new transaction contain an information about the correction on the previous data. To make the ledger light and the network not burden with heavy data, then only the transaction that will be saved on the ledger. That transaction will point at a database that contain passive data. It is analogy with the movement of a batch of coffee on a supply chain based on Blockchain. The property of the batch will stay off grid, but there will be an update on the ledger pointing to that batch.

Another technique for enhancing the integrity of security is by using digital certification. As we can see in Fig. 5, it shows the number of signer public key, batcher public key, transaction ID and header signature. Those are digital certificates for encryption and ensuring the ID of signer, batcher, and nodes. Digital certificate enhances the security on data integrity because hacker cannot easily falsify the information or act as a trusted node to input fraudulent data.

5.3 The Security on Availability

Availability refers to the actual availability of the data even though there is an authentication or access policy to prevent unauthorized person. There are two kinds of data protections on Blockchain, namely chain of block and consensus. Transactions using Blockchain are peer-to-peer, meaning that data (can be messages, money, or important information) can be moved from one user to another without the help of the third party. User no longer depends on one server because all transactions are replicated throughout the network so that avoiding various forms of fraud such as modified data, server down, or hacked accounts. All transactions and data storage are guaranteed to be safe because they are replicated throughout the Blockchain network so that if a hacker wants to change one data, he must get the consensus agreement first to change the same data on all other users' computers at the same time. The consensus is what makes Blockchain different because it is a scheme to protect the system from fraud. Blockchain's consensus mechanism ensures that the next block in a blockchain is the one and only version of the valid transactions, and it keeps powerful adversaries impossible in trying to derail the system and successfully fork the chain. The consensus mechanism solves the inconsistency hurdle and makes it possible to have a reliable distributed system.

The most common consensus mechanism is POW, which is being used by Bitcoin's and Ethereum Classic. As part of the POW system, a nonce is inserted into every block header. A new block is only accepted to be appended to the chain if the hash of the whole block begins with a certain number of zeros (this is known as a "hash puzzle"). Solving the hash puzzle in POW is probabilistic and this lends well to the security assumption of bitcoin system, i.e. most attacks on Bitcoin are unsuccessful if the majority of the miners weighted by hash power are honest in following the protocol, even though there always concern about a vulnerability called 51% attack, when 51% of miner are not honest and try to change the data on their own. The most common criticisms on POW are that it needs a large amount of computational energy and mostly centralized in areas of the world where electricity is cheap or the heat generated as a by-product of mining is needed, China and North of Europe by example. To address these criticisms, a number of alternative consensus protocols (such as Practical PBFT, PoS, Proof of Activity, and PoET) have been proposed [7].

5.4 Blockchain Use-Case on Big Data Security

In this section we use Bettium [13] and TravelChain [14] as an example of the solution of combining Big data and Blockchain for two very different use-case. Bettium is a betting platform of a decentralized marketplace that allows smart sports bets to be made between individuals. The resident experts and the latest analytical data are both on hand for gamers to adopt more calculated strategies for winning. Both the contestants and the teams in a wager who can access Big Data, AI, and many potential analytical tools to enable player's decisions. The platform, rich UX, is applying blockchain technology in ways to give users the benefit of a sophisticated analytical approach in placing bets.

Meanwhile TravelChain describes itself as a decentralized blockchain-based platform which provides accessible and authentic smart data to let business knows their clients better and to help travellers experience the best customer service. When users download the TravelChain application, they will be able to determine how much of their personal data can be accessed by the system. The data that the system can access will be used to figure out the user's personal preferences; like Chinese food, the system will figure that out, and if like horror movies it will see that too. It will be able to put all the things about the user together to generate the best offers for the trip.

Both cases are using Blockchain and Big data for different reasons. Bettium uses Blockchain for enhancing peer-to-peer experience on betting and offering analytics on betting strategy, while TravelChain uses Blockchain for customizing the personal preferences and offering Big data of travel data based on these preferences. However, both applications consider the security features of Blockchain for protecting the value integrity of betting and user's personal data.

A report said that Blockchain system may replace the Society for Worldwide Interbank Financial Telecommunications (SWIFT) system [15] so that more than 100 banks will reduce transfer time for international payments from 3 days to 10 minutes by using the digital currency base of Ripple. Among the banks using Ripple there are such famous banks such as UBS, Unicredit, BBVA, Credit Agricole, and Mizuho. They must consider the security features of Blockchain in deep before deciding that move especially after SWIFT heist occured on Central Bank of Bangladesh in February 2016.

In this context, security is the terminology that is generally called as information security (INFOSEC). On the other hand, veracity is a special characteristic of Big Data. Blockchain with encryption and hash chaining concept is already attracted security industry, and their implementation on improving data veracity via crowdsourcing and provenance methods attracts Big Data industry.

6 The Blockchain's Role in Improving Data Veracity

The defining factors of data collection on Big Data are volume, velocity, variety, and veracity. As has been mentioned at the beginning, volume and velocity characteristics state how much and how soon the data is generated. Variety characteristic states the condition of the data is structured or not, while veracity characteristic speaks about the level of trust in data validity. This level of veracity is where Blockchain can play its role. The final feature of the data where the data is recorded cannot be changed without the record, and make the attacks on data integrity can be avoided. However, Blockchain can only maintain the validity of input devices not the veracity of the data, so that if the garbage data is entered, the junk that will be stored.

Berti-Equille and Borge-Holthoefer presented how to achieve veracity of data based on trust discovery and modelling misinformation [16]. To discover the truth from a structured data, there are many methods like agreement-based method that measures source reputation via hubs and authorities, page ranks, and source ranks, and then source-claim iterative models to fine-grained classification. Another methods are MAP-Estimation-based, Analytical, and Bayesian Methods. Truth discovery from the extracted information came from semi or unstructured data, can be done by using knowledge-based trust and slot filling validation. Another advance methods like evolving truth, truth finding from crowdsourced data, long-tail phenomenon, and approximation of truth existence method can also be used.

Understanding how misinformation spreads is another important thing to achieve veracity of the data. The theory of rumour spreading and its dynamics, information

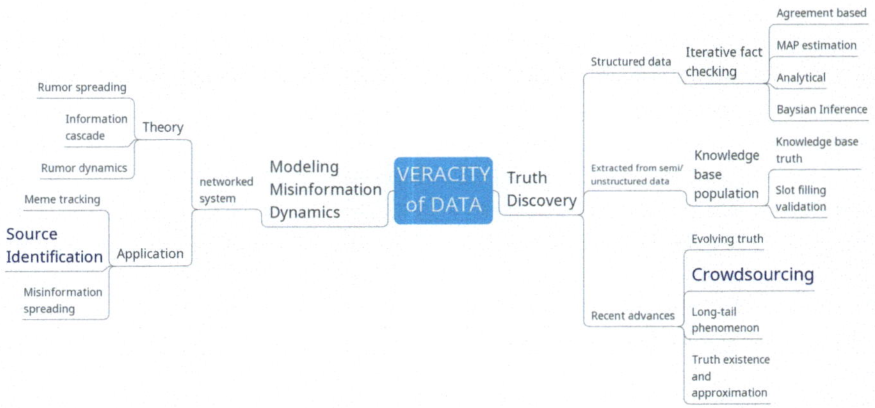

Fig. 7 Veracity of Data [16]

cascading, and how to track meme and ID the source are another methods to understand how misinformation spreads. Figure 7 above shows the map to achieve veracity of data.

TrueCaller is a crowdsourcing example to help user in avoiding disturbing calls, by collecting report of it. Crowdsourcing and source identification are applications that can be provided by Blockchain. Some cases using token as a tool to improve the quality of the input data. Startup like Hara offers this concept of validation by providing a token of appreciation for inputting and validating the data. Another way is to check the ID of data retrieval and measure the level of data truth based on other data. A token-based validation method already proposed to Ministry of Transportation Republic of Indonesia, to validate and reward the data input of passengers on bus transportation mode. On recent development, there are new efforts to enhance data provenance by using decentralised ID based on Blockchain. This method will ensure the ID of data source and enhance data veracity. All of these methods will be explained further below.

6.1 HARA Token

Established in 2003 as Mediatrac, Dattabot's origins were "Big Data" and content acquisition from online and offline sources. Dattabot is focused on solving large societal problems by developing HARA's decentralized data exchange and ecosystem which will support better data-driven decisions for data buyers and incentivize data providers to share their data. The value of data comes from its accuracy. The moment new data enters the system, data qualifiers are incentivized to validate the data which acts as a crowd-sourced indicator of its quality. Overtime, this will help improve the overall robustness of the data and help generate healthy, ongoing demand. Furthermore, data also becomes more valuable when it is processed. So

HARA also provides an 'enriched data' category that allows data buyers to resubmit data that has been analysed and rendered useful while the sharing proceeds with the original data provider(s).

HARA Ecosystem provides decentralized data exchange for data providers to provide their data and the data buyers to access the data. Powered by Blockchain technology, this exchange is traceable, transparent, and secured. Vast amount of data, ranging from farmers identification data, cultivation data, location specific data, ecological data, and market information and transaction data, are collected via a variety of data sources including IoTs and satellites as well as from third-parties such as farmers, governments, scientists, academia, farm-input manufactures, and other entities. The platform will be automated by the smart contracts on HARA's Blockchain and different methods of data submissions, including the HARA suite of mobile applications, IoT devices, satellites imagery data, and more. These data originators are enabled to upload datasets to the data exchange and sell their data to other third-party data buyers via the decentralized token based data exchange [17]. This token-based concept will ensure the veracity of data because there always be a person to update the data and get rewards for it (Fig. 8).

Data providers include all individual data contributors, data companies, cooperatives, Non Government Organizations (NGOs), field agents, and governments. They can use the data exchange to assess the quality of their data and monetize it by exchanging it for tokens. Data qualifiers add value to the ecosystem by providing

Fig. 8 HARA Ecosystem [17]

verification in the form of PoW. They acts as a crowd-source indicator of data quality, which overtime, will help improve the overall robustness of the data and help generate healthy, ongoing demand. Data qualifiers will receive tokens based on their efforts related to the tasks they perform in verifying data on the exchange. Data qualifiers can be any HARA token holders. Data buyers include enterprises such as banks, insurance companies, retailers, agriculture input suppliers, NGOs, and government – all the way down to local communities or even individuals.

Companies and institutions that access and process raw data from HARA ecosystem, and resubmit it as 'enriched data'. These can be academic institutions, brands, data analytics, financial technology and agriculture technology companies. They create value-added insights from the raw data and share in the proceeds with the original data providers by submitting the cleansed, organized, and structured data back to the HARA ecosystem. An example is described simply by using smallholder farmer as any data provider. The following diagram illustrates the flow a farmer will follow to add data to his account and other inputs such as GPS data, crop planting suggestions, pest data and other information associated with their farming practice (Fig. 9).

Smallholder farmers use the HARA platform with the expectation of receiving multiple benefits from the various value-added providers in the ecosystem. By being a part of HARA, they gain access to many benefits such as access to precision farming advice catered specifically to their farm to help them produce greater yields per hectare accompanied with instructions and guidance, recommendations regarding the best farm inputs – crop types, seed types, fertilizers, pest controls, irrigation and more that are specific to each farmer's location and environment, access to market information specific to each farm including: local supply chain partners, post-harvest off-takers, traders, and wholesalers, access to loans and other financial

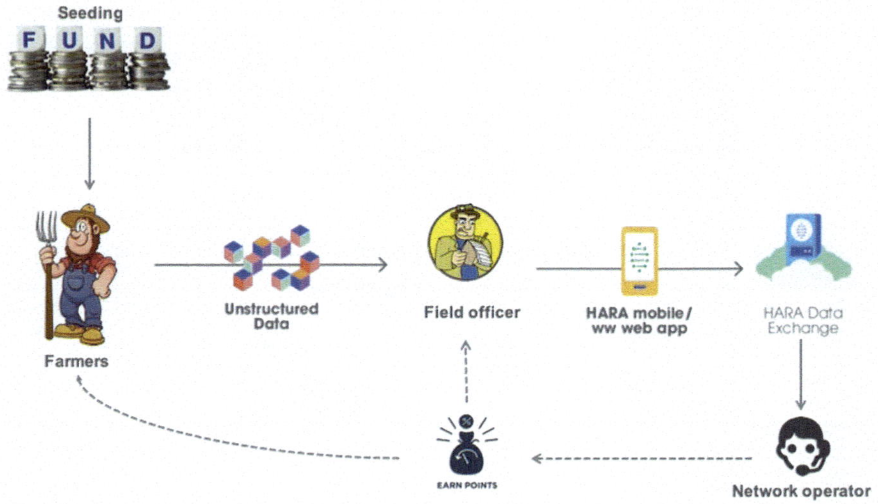

Fig. 9 Loyalty Points To Enhance Data Veracity [17]

services by uploading certified land title document and other associated data, access to other non-banking financial services such as crop/health/life insurance, access to market data related to supply and demand levels and local market prices for each type of crop, and ability to earn loyalty points that can be redeemed for a variety of goods and services provided by HARA partners.

By using this method, HARA can gain trusted data and enrich it by utilizing the field officers with HARA token as the loyalty point. This ecosystem will continue to grow because each stakeholder gets their reward and benefit from the platform based on crowdsourcing method.

6.2 *Passengers Data Case*

On the Ministry of Transportation, there were problems to get valid information of passenger number especially on each bus transported from one terminal to another. Even though they already gave assignment to terminal officer, it still lacked achievement on this task and its data validity. The information needed are total passenger, men, women, and child number. This data can only be achieved at Aidil Fitr, Christmas, and New Year moments because there are special team assigned for the job [18].

Figure below is taken from Information System of Indonesia Transportation and Infrastructure or SIASATI (Sistem Informasi Angkutan dan Sarana Transportasi Indonesia), a web sites hosted by Ministry of Transportation located on http://siasati-dev.dephub.go.id/dashboard/event. The site is only on Bahasa Indonesia language, shows number of passenger on all of transportation mode at Christmas 2018. There are five modes of transportation: land transportation (green line), ferriage (blue line), train (red line), sea transportation (purple line), and air transportation (light brown line). The data is presented from 20 to 31 of December 2018 in the form of line chart and table. As stated before, the Ministry of Transportation still failed to achieve all year data because of the lack of input and the data validity on land-sea-ferriage transportation modes, even though the data from train and air passenger were received (Fig. 10).

To cope with the problem on the land transportation, the proposed solution is token-based achievement control. Using smartphone, the officer gets a targeted number of tokens to be fulfilled each day. The inputs will be compared with inputs from the next terminal. If they carelessly input the data, the system will punish them by reducing their token, vice versa. There must be adjustment calculation too because some passengers are in and out from the bus not inside the terminal.

From the illustration in Fig. 11, the yellow bus is the asset data that moves along the time and updated by many points. What components are updated? For dataset example here are Departure time, Passenger Number, and Arrival Time. The officer at Bus Station A will be the first of the day updater, and then the bus driver updates the dataset based on passenger who gets on and gets off the bus during the travel. The officer of Bus Station B at the next stop will be another updater. An AI can be

MODA ANGKUTAN	20 Des H - 5	21 Des H - 4	22 Des H - 3	23 Des H - 2	24 Des H - 1	25 Des H	26 Des H + 1	27 Des H + 2	28 Des H + 3	29 Des H + 4	30 Des H + 5	31 Des H + 6	01 Jan H + 7	02 Jan H + 8	03 Jan H + 9	04 Jan H + 10	05 Jan H + 11	06 Jan H + 12	TC
JALAN	104.750	117.028	146.651	152.780	133.302	133.363	123.833	121.847	122.112	161.978	159.251	111.184	103.555	114.924	94.789	55.673	58.243	63.477	2
PENYEBRANGAN	134.910	178.798	203.033	151.128	153.893	149.378	130.407	145.539	148.798	157.663	144.575	94.456	83.304	142.951	104.502	0	0	0	2
KERETA API	290.013	304.101	339.995	343.514	328.377	334.369	319.144	312.714	307.851	339.505	349.697	327.904	332.317	293.640	267.207	263.347	284.048	304.650	5
LAUT	161.576	65.504	74.692	73.975	66.092	51.148	52.195	48.415	52.161	53.039	52.046	39.544	39.284	49.848	45.126	49.563	48.859	56.264	1
UDARA	282.619	305.700	306.111	274.679	229.486	254.885	258.646	236.260	247.818	241.766	225.931	180.515	230.646	254.276	244.036	242.051	250.272	264.759	4
TOTAL NASIONAL	973.868	971.131	1.070.482	996.076	911.150	923.143	884.225	864.775	878.740	953.951	931.500	753.603	789.106	855.639	755.660	610.634	641.422	689.150	15

DATA PER Minggu, 17 Februari 2019

Fig. 10 The number of passenger on Christmas for all transportation modes in year 2018 (Source: The Ministry of Transportation – Republic of Indonesia)

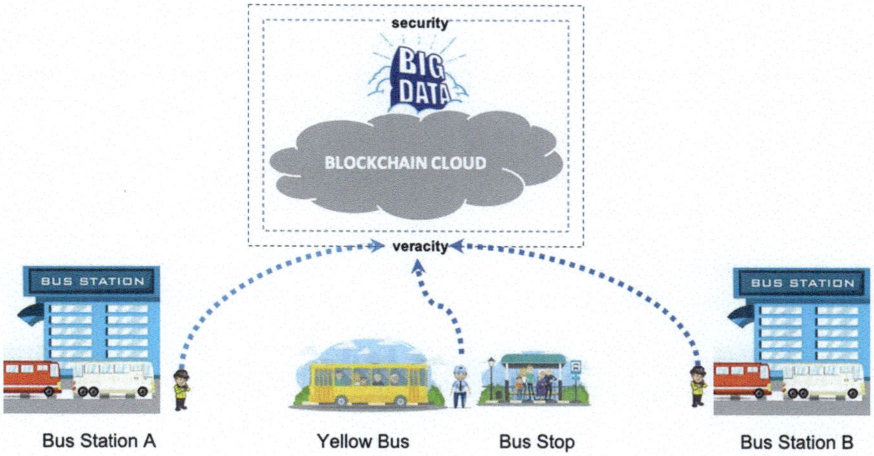

Fig. 11 Illustration of the data movement in Blockchain

used for comparing all the update so that the system will know which data is the correct one and gives reward to the updater.

With adjustment and feedback, this concept will enhance the veracity of Big Data regarding bus passengers on land transportation. This validity is very important to prepare and plan the development of facilities and infrastructure of land transportation, especially intercity transportation. For inner-city transportation, the solution will need IoT tools because passenger gets on and gets off more frequently.

6.3 *Decentralized Identity*

Another approach to enhance data veracity is by using data ID for data provenance. Provenance describes the origin, derivation, and ownership of data products so that it enhances their trustworthiness. However, data with proven ID will have a privacy risk and may contain private information, thus, it may need to be sanitized before released. Parisa Kianmajd [19] presented a model-based diagnosis framework to pinpoint conflict among the policies and showed how it can be used to find conflicts between two sets of publication and customization when sanitizing a provenance graph using ProPub [20]. In this paper we use ID terminology as the abbreviation of word "Identity", except in a special case such as Decentralized Identifiers (DIDs) where ID refers to "Identifier".

Digital identifications have the potential to solve many of the problems associated with physical ID such as government-issued cards or papers. Countries like Estonia and Pakistan with their national ID programs, and India with its Aadhaar platform, have successfully established large-scale national ID systems to streamline the service access for their citizenry with a single identity. United Nations High

Commissioner for Refugees (UNHCR) also creates biometrically-enabled digital IDs for hundreds of thousands of undocumented refugees to allow them having better access to the aid programs and supports. They empower a user to have control over their ID claims without relying on physical document that could easily be lost, damaged, or controlled by a malevolent party.

However, digital IDs must provide security through mechanisms like multi-factor authentication, and have increased resistance to forgery, misuse, or theft. They can also enhance the privacy through the contextual and consent-based disclosure of personal information when others wish to validate claims, such as license status or if a person has reached age of majority, without having to disclose irrelevant details like full date of birth or home address. One of the main risks of traditional digital IDs is in the centralized storage of personally identifying information that possible to become a target of data breaches. News reports of data breaches compromising the privacy and personal information of millions of people are often publicized to harm public confidence in the ability of companies and government to protect their data. Not only that, having a master copy of the data stored in a single location as single point of failure opens up the chances of data corruption or data loss. This makes data security, storage, and privacy protection mechanisms are the key points when evaluating the ID providers. Filling the gap of securing the IDs, Blockchain finds a place in many new digital ID services on the market. Its various cryptographic security mechanisms, data immutability, and decentralized network architectures can be leveraged for the design of secure, user-owned, attestation-based digital ID solutions.

Decentralized Identifiers (DIDs) are a new type of identifier for verifiable, self-sovereign digital identity. DIDs are fully under the control of the DID subject, independent from any centralized registry, identity provider, or certificate authority. Each DID Document may contain at least three things, namely proof purposes, verification methods, and service endpoints. Proof purposes are combined with verification methods to provide mechanisms for proving things. For example, a DID document can specify that a particular verification method, such as a cryptographic public key or pseudonymous biometric protocol, can be used to verify a proof that was created for the purpose of authentication. Service endpoints enable trusted interactions with the DID controller.

This specification of W3C-DID [21] was published by the W3C Credentials Community Group, a group which mission is to explore the creation, storage, presentation, verification, and user control of credentials, focus on a verifiable credential (a set of claims) created by an issuer about a subject—a person, group, or thing—and seek solutions inclusive of approaches such as self-sovereign identity, presentation of proofs by the bearer, data minimization, and centralized, federated, and decentralized registry and identity systems.

A DID is like a Uniform Resource Locator (URL), it can be resolved or dereferenced to a standard resource describing the entity and unlike a URL. The DID Document typically contains cryptographic material that enables authentication of an entity associated with the DID. Figure 12 is an example of a DID Document that describes the DID based on example ID 'did:example:123456789abcdefghi'. The

```
{
  "@context": "https://w3id.org/did/v1",
  "id": "did:example:123456789abcdefghi",
  "authentication": [{
    // this key can be used to authenticate as did:...fghi
    "id": "did:example:123456789abcdefghi#keys-1",
    "type": "RsaVerificationKey2018",
    "controller": "did:example:123456789abcdefghi",
    "publicKeyPem": "-----BEGIN PUBLIC KEY...END PUBLIC KEY-----\r\n"
  }],
  "service": [{
    "type": "ExampleService",
    "serviceEndpoint": "https://example.com/endpoint/8377464"
  }]
}
```

Fig. 12 Minimal Self-Managed DID Document [21]

'example' column is the method like Sovrin, Civic, etc. The '123456789abcdefghi' is the ID number but it usually uses base58 string encoder. The standard assumes that the entity that controls the private keys for this identifier is authoritative for the DID Document.

The following is an example of ID Provider based on Blockchain based on document from Blockchain Learning Group and Blockscale Solution [22]. The report compares ID providers, platform, and national ID. It focuses on the issues of their purpose, the solution, unique features, status, position on self-sovereign ID, General Data Protection Regulation (GDPR) compliant, Blockchain based, private storage mechanism, recovery mechanism, problem and mitigation, interoperability, attributes attestation, business model, liability framework, attributes sharing, user privacy, claims made and verifying mechanism, application, and so on. Civic, uPort, Sovrin, and ShoCard are decentralised ID based on Blockchain.

Civic is a Blockchain-powered identity company that provides businesses and individuals with tools to control and protect their identity. The Civic Secure ID enables personal verified information to be stored on user mobile device using bank level encryption and biometric locks. Users safely access partner websites and services using the Civic Secure ID. Civics' Secure ID Platform (SIP) enables businesses to leverage the Civic Secure ID via the Civic App, Integration Portal, Civic Web Connect, and Civic App Connect to perform login, age verification, and know your customer verification activities. uPort is a platform for self-sovereign identity and user-centric data management, anchored in Ethereum Blockchain. It provides a mobile application which enables users to create a digital identity in a few simple clicks. From there, users can start interacting with uPort compliant services. For example, in the city of Zug, Switzerland, a citizen can receive a credential verifying that he/she is a resident, and uses that credential via uPort for multiple services like eVoting or eBiking. uPort also provides a rich set of tools and Software Development Kits (SDKs) that developers can use to add uPort capabilities into their applications or solutions, guaranteeing they will deliver services their end users can trust.

The Sovrin Network is a public service utility enabling self-sovereign identity on the Internet. This decentralized identity network offers enterprises and developers free, open source code to create private and secure data management solutions that run on Sovrin's identity network. The Sovrin Network allows individuals to collect, share, and manage the individual components that make up their identities. ShoCard is a digital identity and authentication platform built on a public blockchain data layer, using public/private key encryption and data hashing to safely store and exchange identity data, which includes biometrics such as fingerprint, facial, iris and voice. ShoCard's approach to identity is different than the existing solutions in that the user owns and carries his/her own data within his/her mobile application and is the sole person who decides with whom to share it with and which pieces of identification to share. The blockchain is then used to validate that information and confirm other third parties who have definitively certified the identity of the user.

The main problems of identification method are the registration and the validation. Physical ID card costs a lot of money and time for them but it is state responsibility to serve their citizen. However for digital purposes, such physical ID often has a limited function and easily faked. Therefore, Blockchain decentralised identity will ensure the provenance of data by using various type of authentication such as biometric, Personal Identity Number (PIN), or password.

6.4 The Comparison among Methods

Based on Fig. 7, the methods presented in this paper are crowdsourcing and decentralized identity. Using decentralized identifier will enhance data provenance even though they need to be sanitized at first to comply with the data privacy protection rules. However, the standard of decentralized identifier is not taking place yet to be an international standard so all DID applications can talk to each other.

The similarity among TrueCaller, HARA Token, and Ministry of Transportation's case are crowdsourcing methods. The etymology of the word "crowdsourcing" comes from a combination of 'crowd' and 'outsourcing'. In layman's terms, crowdsourcing is the process of making content openly available to the public for use and verification. Crowdsourcing is the practice of engaging a 'crowd' or group for a common goal - often innovation, problem solving, or efficiency. It is now easier than ever for individuals to collectively contribute - whether with ideas, time, expertise, or funds - to a project or cause. The underlying idea is to divide a project or task into smaller pieces and assign it to the masses. The major advantage is that one has the resources of tens of thousands of people at one's disposal to accomplish the task. Owing to the large number of participants, the quality of resulting content is far more superior, especially in terms of ideas and diversity [23].

The differences are their business process. TrueCaller using vast amount of contact databases collected from their user and allow their user to report inappropriate number so that another user can benefit from it. Sample given, a number using by trickster will be reported and show when that number calls another user as a trickster

Table 1 The comparison results among methods

Comparution	TrueCaller	HARA Token	Data Passenger Case	Decentralized ID
Using Blockchaln	no	yes	yes	yes
Data Processed	contact number	agriculture data	passenger data	identity claims
Target	filtering irritating caller	data accuracy	data accuracy	sell sovereign identity
Methods	crowdsourcing	token of appreciation	token of appreciation	W3C-DID
Reward for Data Source	none	money	performance index	money
Business Model	freemium	selling data analytics	government report	consortium fee
Big Data Aspect	veracity	veracity	veracity	provenance

number. The more user reports the number, the more reliable is the data. TrueCaller then asks a donation or offers premium services such as online information of contact and trusted number for small amount of money.

The similarities and the differences can be seen on Table 1. The question still remains, is TrueCaller will be a better solution if using Blockchain? The answer can be yes or no. If TrueCaller will change their business model by using token as a tool for appreciating the value, Public Blockchain will be a good platform for it. However, actually TrueCaller is not a solution based on multi stakeholder problem that needs a common data set, so that centralised database is enough for now.

The HARA Token and the Ministry of Transportation's case use different approach of crowdsourcing. HARA Token offers a token for inputting and validating data which can be changed to fiat money. By having highly data veracity, HARA can offer the analytics for the business and gains margin. Ministry of Transportation's case on collecting data passenger uses the token for key performance index of their officers. The officers have targets to be fulfilled and therefore their working performance will be depended on it. On the other hand, the Ministry can give them rewards for achieving certain targets to boost their morale. Blockchain-based decentralized ID can enhance crowdsourcing method over data provenance. The data source can be traced because the ID of data source can be trusted. The method will enhance the data veracity of Big Data.

Furthermore, Blockchain can be used to enhance security and veracity of Ministry of Transportation's SIASATI Big Data. Back to Fig. 10, one solution is already proposed to cope with the problem of data input and its validation. However, the application itself is still built on centralised database while after the assessment using BIAF, this application is passed for Blockchain solution. Using Blockchain, the data that was previously collected on the Ministry will be a common dataset that can be used by other stakeholders (Fig. 13).

Before implementing Blockchain, the data is centralised on Ministry's premise(s). Centralization increases the security risk because if the datacentre is compromised, the system will be breached and all data will be possibly stolen, or lost its integrity.

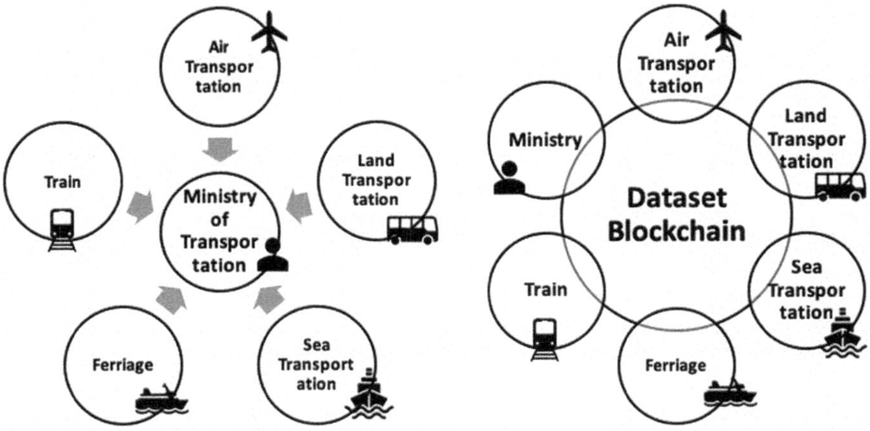

Fig. 13 The view of the system before and after the Blockchain implementation

Besides security consideration, there is usage consideration, too because only the Ministry of Transportation which can use the data. After the implementation, there will be a common data set that is agreed by each stakeholder to be shared and updated. Each party can read and write the data based on the permission rule and each data source has a responsibility on the data veracity and the data provenance using variety of methods such as crowdsourcing and decentralized ID. From security point of view, an attack on one premise cannot change the data because the data set is already protected with a robust encryption method and a consensus. To increase the privacy and the security, Private Blockchain is more recommended to be used than Public one.

7 Concluding Remarks

7.1 Conclusions

Blockchain is not a solution for all problem. Understanding of the problem and the technology are needed to select the best solution for it. For a specific problem, Blockchain can enhance the security of Big Data by strengthening the security of the data storage. Blockchain enhances data confidentiality by data encryption, enhances data integrity by using digital certificate and chaining the block using hash of previous block, and enhances data availability using peer-to-peer transmission, distributed nodes, and consensus method.

Blockchain can also enhance the performance of Big Data Analytic by providing a better data veracity from token-based validation and ID decentralization. Token-based validation enhances truth discovery by crowdsourcing method using token of appreciation concept as reward or just performance measuring. Data veracity can be enhanced more by using decentralized identifier to prove the identity of data source. However, it must be sanitized first to comply privacy data protection rules.

7.2 *Further Works*

Blockchain is a disrupting technology, still a long way to be mature. There are many research and development efforts around the world to take the opportunity of this technology as well as the effort to make standardization of Blockchain. Further works to be done as follows:

- using feature off-ledger and on-ledger for storing massive data in quantity
- true implementation of sanitized DID on Big Data use-case and comparing its veracity level after implementation
- using Blockchain as a tool for another enhancement effort on Big Data Analytics industry

References

1. IBM (2017) Big data analytics. https://www.ibm.com/analytics/hadoop/big-data-analytics. Accessed 11 Feb 2019
2. El-Seoud E-S, Abdelfattah M (2017) Big data and cloud computing: trends and challenges. iJIM 11(2):2017
3. Leskin P (2018) The 21 scariest data breaches of 2018. https://www.businessinsider.sg/data-hacks-breaches-biggest-of-2018-2018-12/?r=US&IR=T. Accessed 11 Feb 2019
4. Puget JP (2015) Optimization is ready for big data: part 4, veracity. https://www.ibm.com/developerworks/community/blogs/jfp/entry/optimization_is_ready_for_big_data_part_4_veracity?lang=en. Accessed 11 Feb 2018
5. Garrick H, Michels R (2017) Global Blockchain benchmarking study. University of Cambridge, Cambridge, p 20
6. Wibowo S, Ery PH (2018) Jangan Pakai Blockchain! https://inet.detik.com/cyberlife/d-4053073/jangan-pakai-blockchain. Accessed 11 Feb 2019
7. Nikolai K (2018) From PoS to dBFT: a brief review of consensus protocols. https://cointelegraph.com/news/from-pos-to-dbft-a-brief-review-of-consensus-protocols/amp. Accessed 8 Dec 2018
8. The Hyperledger Architecture Working Group (WG). Hyperledger Architecture, Volume 1: Introduction to Hyperledger Business Blockchain Design Philosophy and Consensus
9. Satriyo W, Ery PH (2018) Blockchain Implementation Assessment Framework, Case Study of IoT LPWA Licensing in Indonesia, 2018 International Conference on ICT for Smart Society. ISBN 978–1–5386-6589-3/18
10. Leavitt HJ (1964) Applied organization change in industry: structural, technical and human approaches. In: Cooper W, Leavitt HJ, Shelly MWI (eds) New perspectives in organization research. John Wiley, New York, pp 55–71
11. Es-Samaali H, Outchakoucht A, Leroy JP (2017) A Blockchain-based Access Control for Big data. Int J Comput Netw Commun Secur 5(7), July 2017, 137–147, E-ISSN 2308–9830 (Online)/ISSN 2410–0595 (Print)

12. Ayyub A, Mazhar AM (2018) Confidentiality in Blockchain. Int J Eng Sci Invent (IJESI), ISSN (Online): 2319–6734, ISSN (Print): 2319–6726. www.ijesi.org ||Volume 7 Issue 1|| January 2018 || PP 50–52
13. Garg P (2018) Bettium: the betting platform will operate at intersection of big data and Blockchain. https://btcmanager.com/bettium-the-betting-platform-will-operate-at-intersection-of-big-data-and-blockchain/?q=/bettium-the-betting-platform-will-operate-at-intersection-of-big-data-and-blockchain/&. Accessed 12 Feb 2019
14. Guest P (2019) How big data and Blockchain technology will change tourism forever. https://btcmanager.com/big-data-blockchain-tech-will-change-tourism-forever/?q=/big-data-blockchain-tech-will-change-tourism-forever/&. Accessed 12 Feb 2019
15. Report.az (2017) Banks change SWIFT to blockchain system. https://report.az/en/finance/banks-change-swift-to-blockchain-system/. Accessed 12 Feb 2019
16. Berti-Equille L, Borge-Holthoefer J (2015) Veracity of big data. https://hal.inria.fr/hal-01856326/document.
17. Hara (2018). Blockchain for better decisions: a global & transparent Blockchain-based data exchange. Whitepaper. https://haratoken.io/doc/ HARA_Token_White_Paper_v20181206.pdf. Accessed 14 Feb 2019
18. Kemenhub P (2018) Metode Implementasi Blockchain Dalam Aplikasi Siasati, Academic Study, Unpublished
19. Parisa K (2017) Protecting data privacy in the presence of data provenance, Dissertation, University of California Davis
20. Dey SC, Zinn D, Ludäscher B (2011) ProPub: towards a declarative approach for publishing customized, policy-aware provenance. 225–243. https://doi.org/10.1007/978-3-642-22351-8_13
21. W3C Community Group (2019) Decentralized Identifiers (DIDs) v0.11. https://w3c-ccg.github.io/did-spec/. Accessed 11 Feb 2019
22. Chami A (2019) Digital ID – a report on the digital identity landscape & providers. Blockchain Learning Group, Inc.& Blockscale Solutions. https://blockchainlearninggroup.com. Accessed 11 Feb 2019
23. Agarwal B, Ravikumar A, Saha S (2016) A novel approach to big data veracity using crowdsourcing techniques and Bayesian predictors. https://doi.org/10.1145/2998476.2998498. https://dl.acm.org/citation.cfm?id=2998498. Accessed 11 Feb 2019

Ir. Satriyo Wibowo, S.T., M.B.A., M.H., IPM graduated his Bachelor Degree from ITB (Institute of Technology Bandung) majoring in Electrical Engineering in 2003, and then took two master degrees, namely Master of Business Administration majoring in Business Leadership from ITB in 2012 and Master Degree in Law from Gajah Mada University majoring in Business Law in 2015. Currently he is assisting the Ministry of Communication and Informatics of Indonesia, as an expert in internet numbering, IoT security, digital ID, and Blockchain. He is also registered as a cybersecurity governance expert on National Resilience Council (Dewan Ketahanan Nasional), National Committee on Economy and Industry (Komite Ekonomi dan Industri Nasional), National Energy Council (Dewan Energi Nasional), and National Cyber and Encryption Agency (Badan Siber dan Sandi Negara). His research focuses are cybersecurity governance and workforce, IPv6, IoT, cyber jurisdiction, Blockchain, and Common Criteria.

Colonel (Electronic) Dr. Ir. Arwin Datumaya Wahyudi Sumari, S.T., M.T, IPM was inaugurated as 2nd Lieutenant Officer of Electronics Corps of the Indonesian Air Force from Indonesian Air Force Academy, Yogyakarta, and achieved Adi Makayasa Medal as Best of the Best Graduate. He received S.T. degree in Artificial Neural Network, M.T. degree in Multi Agent System, and Doctor in Cognitive Artificial Intelligence from Institut Teknologi Bandung (ITB), Indonesia in April 1996, March 2008, and July 2010. All degrees were achieved with Cum Laude. His research interests are Cognitive Artificial Intelligence, multi agent systems, and cybersecurity. He is a Senior Researcher at CAIRG, ITB. He has written more than 200 technical and general papers published internationally and nationally, and is also Steering Committee in several International Conferences. Colonel Dr. Arwin Sumari holds several professional certifications and he currently is a Senior Electrical Engineer Officer at Abdulrachman Saleh Air Force Base, 2 Operation Command, Indonesian Air Force, Malang, East Java, Indonesia. He is also Adjunct Professor at State Polytechnic of Malang and Assistant Professor at Indonesia Defense University.

Authentication Methodology for Securing Machine-to-Machine Communication in Smart Grid

Zubair Md. Fadlullah and Mostafa M. Fouda

Abstract The bidirectional communication between the smart grid users and utility company is facilitated through Advanced Metering Infrastructure (AMI) comprising numerous smart meters, sensors, and other Internet of Things (IoT) devices by employing Machine-to-Machine (M2M) communication. Triggered by advances in the M2M technologies recently, the smart meters do not require any human intervention to characterize power demand and energy distribution. While there are many challenges in the design of the smart grid communications network, security is a major obstacle in realizing smart grid communication. This is because of the convergence of the advanced IoT and M2M technologies at the smart grid arising many new unforeseen challenges leading to security vulnerabilities and malicious threats. Therefore, practical and lightweight authentication mechanism for fulfilling the specific requirements of the smart grid communication should be carefully taken into consideration and adequate authentication methodology should be developed tailored for the smart grid context. In this vein, in this chapter, we first overview the M2M communication framework in the smart grid system and highlight its shortcomings including security vulnerabilities such as communication trust, and privacy issues. In order to deal with the security concerns, a lightweight message authentication method is presented to carry out mutual authentication among the smart meters distributed at the various hierarchical networks of the smart grid. The adopted lightweight authentication method is based on Diffie-Hellman key exchange protocol. A cryptographic analysis of the adopted authentication method demonstrates its ability to satisfy the desirable security demands of the smart grid communications. Simulation results are also provided to demonstrate the viability of the adopted

Z. Md. Fadlullah (✉)
Computer Science Department, Lakehead University, Thunder Bay, ON, Canada

Thunder Bay Regional Health Research Institute, Thunder Bay, ON, Canada
e-mail: zubair.fadlullah@lakeheadu.ca

M. M. Fouda
College of Engineering, Tennessee Tech University, Cookeville, TN, USA

Faculty of Engineering at Shoubra, Benha University, Benha, Egypt
e-mail: mfouda@ieee.org

Z. Md. Fadlullah, A.-S. Khan Pathan (eds.), *Combating Security Challenges in the Age of Big Data*, Advanced Sciences and Technologies for Security Applications, https://doi.org/10.1007/978-3-030-35642-2_9

authentication method. In addition, the need for developing another specific type of authentication for securing targeted broadcast in the smart grid system is discussed and the applicability of Key Policy Attributed Based Encryption (KP-ABE) is investigated for this purpose. It is shown that the smart grid's control center can employ KP-ABE to broadcast a single, encrypted message to specific groups of recipients whereby each group consists of numerous users. Each user in the targeted group is able to individually exploit the defined key policy to decrypt the broadcasted message. It is demonstrated that in such highly specialized communication scenario, the adopted KP-ABE targeted broadcast methodology is capable of eliminating the need to issue redundant/unicast messages to ensure both communication and computation efficiency while protecting the confidentiality of the exchanged information in the smart grid.

1 Introduction

Human civilization owes a great to the advancement of power generation and distribution technologies. Undoubtedly the proliferation of the power grid transformed our civilization during the last century. On the other hand, in the current century, digitization is again transforming our societies in an unprecedented manner toward a smart society. Indeed, the smart buzzword has engulfed us in all walks of life with a plethora of smart devices and technologies such as smart home, smart industry, smart health, smart driving, and so forth. The current power generation and dissemination grids are almost a century old and they are facing both capacity and scalability issues to meet the growing, dynamic power requirement of the highly connected population in the current Internet age era. Governments of various countries have acknowledged this fact and have commenced the smart grid projects to address this growing concern which may permit intelligent power control and monitoring. Thus, idea of smart grid has been receiving tremendous attention amongst both researchers and utility companies recently. In particular, in the smart grid, advanced technologies like the Internet of Things (IoT) sensing, control, instrumentation, digital communication, and network information, as shown in Fig. 1, are taken into account in addition to the traditional one-way power system engineering. This is done so as to efficiently address a number of major issues which restrict the performance of existing electricity grids including the lack of appropriate demand response, scalability, energy conservation, reduction of carbon emission, and control of energy distribution. Smart grid is regarded as a transforming technology for the energy industry to facilitate bi-directional communication between the electricity company and its customers [1–10]. The two-way communication can take place by means of smart meters deployed at homes, buildings, neighbor sites and control centers. Smart meters, along with various power instrumentation and monitoring sensors, form the Advanced Metering Infrastructure (AMI) as shown in Fig. 2. The smart meters "talk" to one another using the Machine-to-Machine (M2M) communication

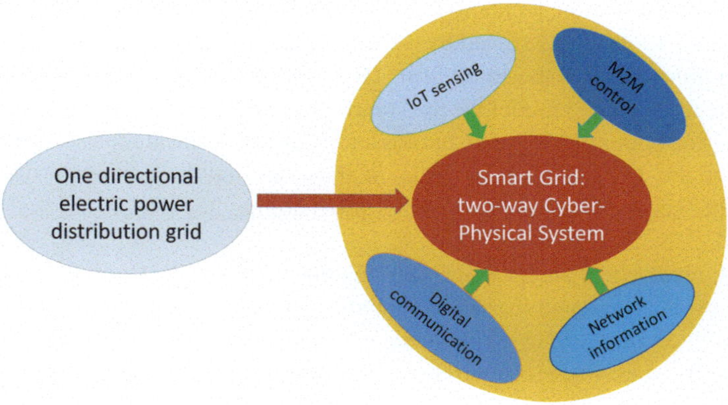

Fig. 1 The evolution of the modern smart grid as a two-way Cyber Physical System (CPS) from the traditional one way power grid system. The enabling technologies are shown that include M2M control, network information, digital communication, and IoT sensing

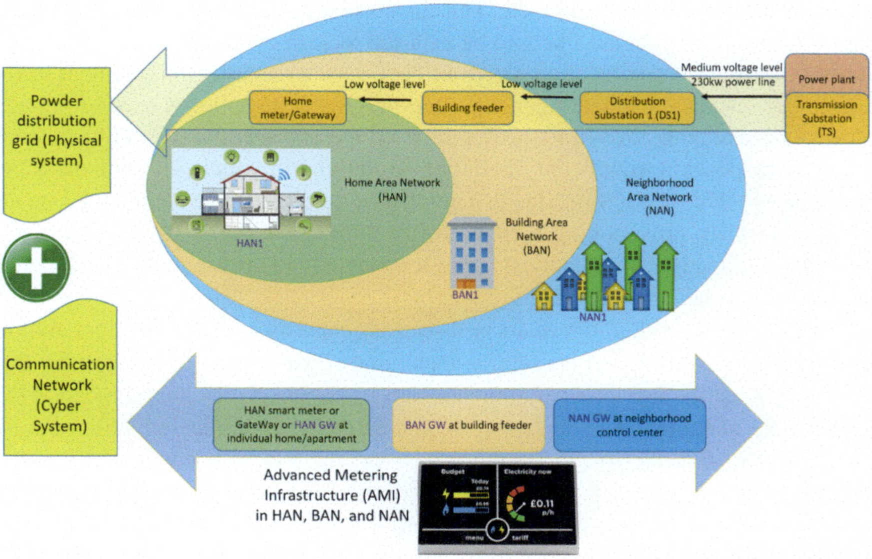

Fig. 2 Considered smart grid architecture. NAN, BAN, and HAN refer to Neighborhood Area Network, Building Area Network, and Home Area Network, respectively

paradigm using wireless communication channels. M2M is at the core of the Internet of Things (IoT) based Cyber Physical Systems (CPSs) like the smart grid which enable the smart meters and sensors to communicate the data it records on power quality, power consumption, energy demand, temperature, sleep/wake up time of devices, interference, coverage area, interoperability, and so forth. The M2M communication architecture allows the heterogeneous devices in the smart grid to interact with one another without the human intervention. In this chapter, we review how existing communication technologies like IEEE 802.11 (WiFi), IEEE 802.15.4 (ZigBee), Bluetooth, and so on address the bandwidth and delay requirements of the M2M communication in the smart grid. Then, we provide a detailed Machine-Type Communication (MTC) model for the smart grid and verify the effectiveness of different adopted communication technologies. Next, we delineate the security vulnerabilities of the M2M communication in the considered smart grid system and discuss the need of adequate security provisioning technique tailored for the smart grid.

Indeed, researchers and practitioners dealing with various smart grid pilot projects have often articulated the need to seamlessly place a security infrastructure at the AMI of the smart grid [3]. This is because many smart grid initiatives are widely employing the state-of-the-art wireless communication networks to connect a multitude of sensors and smart meters to link customers with the utility company so that they can exchange information both ways. As the readers may guess, the first idea would be to use the existing Internet infrastructure to provide the connectivity in the smart grid system. In the recent years, however, there have been a plethora of attacks on the Internet such as the Distributed Denial of Service (DDoS) attacks, spoofing, privacy leakage, worm infection, etc. This means that exchanging grid related information over the Internet are likely to be exposed to such cyber-attacks also [11]. In addition, various communication networks have various security enforcement techniques. For instance, the heterogeneity of cryptographic techniques used to protect information exchange in WiFi is different from 4G and beyond 4G cellular communications poses a significant challenge to determine which security measure should be adopted. In addition, due to a number of unique requirements of the smart grid, the security provisioning technique like authentication of the smart meters need to be lightweight and practical without incurring much computation and communication overhead.

Traditionally speaking, the smart grid security consists of authentication, authorization, access control, and encryption techniques. These techniques can offer the basic yet critical security provisioning demands in the smart grid communication by securing the customer data, incentive plans, and other confidential information. Therefore, the first step is to design a lightweight and efficient authentication method for the smart meters so that only legitimate users are serviced. Beyond the authentication process, there are other security needs which should be taken into account. For instance, the smart grid's control center, from time to time, may disseminate some sensitive information to a specific group of users, and the sensitive information should be transmitted in a secure manner. If the control center carries out unicast transmission to each user individually, the communication cost will be quite high due to the large pool of smart grid users/customers. Furthermore, this issue may become even more complicated when the control center generates different service types for different groups of users. Hence, how to securely and efficiently disseminate sensitive information to the targeted user groups is a significant challenge. Intuitively, to address this issue, conventional broadcast encryption techniques may be employed whereby the control center initially chooses a set of privileged users, and then transmits a single encrypted message to those users, and later only those privileged users are able to decrypt and access the message. However, when the set of privileged users is dynamically changed, the traditional techniques are rendered impractical. In this chapter, towards secure and flexible targeted broadcast in the smart grid, we consider a new targeted broadcast mechanism with Key Policy Attribute-Based Encryption (KP-ABE) [12], where the control center is permitted to dynamically choose user-groups based on their attributes. In this methodology, each user in the smart grid is associated with a set of attributes based on which the user's decryption key is derived. Upon receiving an encrypted message transmitted by the control center along with an access policy based on these attributes, the appropriate users, who satisfy the access policy, alone, are able to decrypt the message.

The remainder of the chapter is organized as follows. Section 2 discusses the background and related research work. Our considered smart grid communication architecture is described in Sect. 3. Based on this, a number of wireless communication technologies are discussed and an appropriate technology is adopted for M2M communication in the smart grid in Sect. 4. Next, in Sect. 5, the security challenges in the smart grid M2M communication are discussed and a lightweight authentication scheme to address these challenges is presented. In Sect. 6, a special scenario involving targeted broadcast in the smart grid is discussed and a Key Policy Attribute Based Encryption (KP-ABE) methodology is presented to secure such targeted broadcast. Finally, concluding remarks are provided in Sect. 7.

2 Background and Related Research

The IEEE P2030 led the smart grid standardization through three task forces for power engineering, information technology, and communication technology with a substantial overlap between the latter two task forces [13]. The communication technology work group provided broad directives on smart grid design standards focusing on choosing appropriate M2M communication protocol. The technology giants like Verizon Wireless and Qualcomm joined hands to bring M2M capabilities to the smart grid by facilitating technologies to wirelessly connect utility companies to grid elements like circuit breakers, transformers, substation equipment, and so forth. The wireless network based machine-centric communication allows the utility company to develop interactive services. The M2M market is expected to grow at a high rate due to increasing investment by both government and private enterprises to adopt M2M services in the coming years [14]. The work in [14] explored M2M communication applications and scenarios such as medicine, transportation, environmental monitoring and smart grid, which are growing and leading the way to new business cases. The work revealed the practical requirements and threats of M2M application scenarios and pointed out the delay sensitive requirement of M2M applications. Due to such unpredictable connectivity to the core network and the demand for high configurability and flexibility of M2M devices, an earlier research work in [15] attempted to identify security threats against M2M communications.

However, the exact technologies upon which M2M communications are based were not taken into account in that research work.

The research work conducted in [16] analyzed the main performances of Internet Protocol version 6 or IPv6 over Low Power Personal Areal Network (6LoWPAN) in order to evaluate their applicability in supporting the smart grid operations. Their research showed the inter-cluster collision problem arising in such a short range wireless network. However, the work did not investigate the exact smart grid applications which may be benefited from the deployment of such wireless personal area networks in the smart grid home area network level. A number of challenges in the design of the home M2M network were, on the other hand, discussed in [17] that demonstrated that the home networks require effective M2M gateways to carry out information exchange between the numerous M2M devices and to provide a link to the smart grid core network or backhaul. While the backhaul connectivity could be implemented by optical fiber, cable, DSL, Ethernet, or even cellular links, the work suggested that it is also critical to select adequate network protocols to enable M2M devices to communicate in the home networks of the smart grid. As the home area network technology, the work investigated IEEE 802.15.4 (ZigBee/6LoWPAN), IEEE 802.11 (WiFi), and Bluetooth protocols albeit without the consideration of the impact of selecting these technologies in the specific case of smart grid M2M communications in the home area network. In the early part of this chapter, we focus on our chapter on investigating the appropriate M2M communication technology suited for the residential networks belonging to a typical smart grid. Then, we discuss the security issues in the M2M based smart grid communication.

A number of research works addressed the basic security issues of the smart grid communication. For instance, the work in [18] considered power system communication and digital security issues as critical components of the smart grid. The work indicated that several cyber security issues need to be addressed for smart grid communication, and particularly focused on the impact of merging the Supervisory Control and Data Acquisition/Energy Management System (SCADA/EMS) with information technology networks. Such fusion of the physical and cyber elements of the smart grid were pointed out to lead to substantial security threats. Additionally, it was indicated in [18] that broadband Internet technologies could allow intruders to access smart meters and even the SCADA by which they may collect critical information such as the metering data, price information, special offers, and so on. Thus, the privacy protection was identified to be a key security concern in the smart grid. Furthermore, in [19], the stringent security demands of smart grid systems were discussed. For instance, a strong authentication technique was pointed out to be a must for all heterogeneous user devices belonging to the smart grid system. However, this raises interoperability and scalability issues. This is because the strongest authentication methods may not necessarily be the quickest in the smart grid due to the large number of users and things like IoT sensors, machines, home appliances, smart meters, phasor units, and so forth. Hence, scalable key and trust management systems, designed to the particular requirements of the utility company and users, is essential from the smart grid communication perspective.

The research work conducted in [20] showed the need for secure aggregation of data gathered from different smart meters. The work presented 4 protocols to securely aggregate data readings from the smart meters: an interactive protocol, Diffie-Hellman Key-exchange based protocol, Diffie-Hellman and Bilienar-map based protocol, and a low-overhead protocol. Only the first approach does not depend on the original Diffie-Hellman key exchange protocol. On the other hand, the remaining approaches all are based on the Diffie-Hellman key exchange protocol or its other variants with relaxed assumptions. However, the work did not take into account smart meter authentication for which Diffie-hellman based approaches could be adopted. Next, in [21], three methods were compared to authenticate demand response messages through smart meters in the smart grid. The methods were called Bins and Balls (BiBa), Hash to Obtain Random Subsets Extension (HORSE), and Elliptic Curve Digital Signature Algorithm (ECDSA). Among these approaches, ECDSA offers stronger security. However, the higher level of security could be obtained with added computational complexity. In this chapter, based on our presented smart grid M2M communication framework, we present a lightweight message authentication method for smart meters while considering the specific needs of the smart grid communication.

Next, we investigate more unified approaches to secure smart grid communications as discussed in [22], which showed that the traditional approach to facilitate smart grid security is not adequate and may leave the British power supply system vulnerable to cyber attacks. Additionally, the study in [22] also demonstrated that the security mechanisms of smart grid are scattered. Even though the smart meter deployment has recently gained much attention, particularly on the safety and privacy, the way to formulate a holistic security framework for the smart grid communication remains unclear. The viability of recent security mechanisms like Attribute Based Encryption (ABE) methodology [12] in the context of smart grid communication, remains to be explored. In the ABE paradigm [12, 23, 24], descriptive attributes and policies (associated with the user) are employed to decrypt encrypted messages. In this paradigm, at first, a centralized entity generates secret keys for the users based on attributes or policies for each of the users. To permit a user to decrypt an encrypted message, a minimum number of attributes needs to be satisfied regarding the encrypted message and the user's private key. A modified version of ABE, referred to as the Ciphertext-Policy ABE (CP-ABE), was envisioned in [24], where the private key of the user is associated with a set of attributes, and the encrypted message defines an access policy over the attributes. In order to decrypt a certain ciphertext, the user is required to ensure that her attributes satisfies the access policy, specified in the ciphertext. How CP-ABE can be exploited for smart grid communication was demonstrated in [25], where an example shows how CP-ABE offers selective access to the user data stored in the smart grid data repositories. Another variant of ABE referred to as the Key-Policy ABE (KP-ABE) [12] labels each encrypted message with a set of attributes. Every private key is associated with an access structure, which defines which type of ciphertexts the key is able to decrypt. This means that the access structure and ciphertexts in KP-ABE are specified by the private key and the attributes set, respectively. It has been demonstrated that a tree access structure can be adopted in KP-ABE, where the internal nodes consist of AND and OR gates, and the leaves denote the attributes of different users. If a set of users satisfies the tree access structure, each user in that set is able to reconstruct the secret. Later in the chapter, we adopt the KP-ABE technique to achieve secure and flexible targeted broadcast, a very specific requirement of the smart grid which cannot be addressed by traditional security methodologies.

3 Considered Architecture for Smart Grid Communications

Before delving into detail of the M2M communication technologies, let us have an overview of our considered smart grid communication framework as depicted in Fig. 2. As shown in the figure, the smart grid power transmission and distribution system and the communication system, representing the physical and cyber parts of the smart grid CPS, are separated into two planes. We first briefly describe the power Distribution Network of the physical plane of the smart grid. The power plant or plants (which can range from traditional fossil fuel based generation to hydroelectric plant to renewable generation sources) are part of an energy market based on an wholesaler-retailer-distributor model. The power generated at the power plant is delivered to the end users or customers through the Transmission Substation (TS) and several Distribution Substations (DSs). The Transmission Substation, located near the power generation plant site, delivers power from the power plant over high voltage transmission lines (usually over 230 kilo volts) to the Distribution Substations. These Distribution Substations are situated in various regions, and they transform the electric power to medium voltage levels. This medium-voltage level power is distributed to the building-feeders. The building feeders convert the medium-voltage level into a lower level to be used by home appliances.

However, if we consider the communication side or the cyber element of the smart grid, the above consideration may not be applicable since the communication links have requirements that differ from those of the power lines. The Transmission Substation and the Control Centers of the Distribution Substations are linked in a meshed topology using optical fiber. The remaining communication topology of the smart grid is split into a number of networks that mimic urban planning which divides a city into a number of wards and neighborhoods, each comprising many buildings, which may have a number of apartments. Motivated by the urban planning, the communication topology of the smart grid considered in this chapter is shown in Fig. 2, which is split into three levels of networks: Neighborhood Area Network (NAN), Building Area Network (BAN), and Home Area Network (HAN). To keep things simple, consider that each Distribution Substation covers only one neighborhood zone. Each NAN can be considered to be composed of a number of BANs. On the other hand, each BAN contains a number of apartments. In Fig. 1, the apartments are shown to consist of their respective local area networks, each of which is denoted as a HAN. Furthermore, the smart meters deployed in the smart grid architecture is part of the earlier mentioned AMI for enabling the automated, two-way communication between the utility meter and the utility company. The smart meters are equipped with cyber physical interfaces, i.e., the communication gateway and energy dashboard. The smart meters deployed in NAN, BAN, and HAN are referred to as NAN GW (GateWay), BAN GW, and HAN GW, respectively. Additionally, it is also worth noting that based on the existing standards of smart grid, Internet Protocol (IP) based communications network is preferred to allow virtually effortless inter-connections with HANs, BANs and NANs. Below, we describe the smart grid communication networks.

3.1 Neighborhood Area Network – NAN

A NAN is a localized or regional network of the considered smart grid communication framework. A NAN consists of one or more cellular (4G/4G+) base stations and a number of BANs. It is worth stressing that the cellular framework used for smart grid communications could be physically or virtually separated from the existing Internet access networks and core networks to avoid possible security and reliability issues. This is not a must but a recommended assumption. Also, note that other modes of communications such as satellite connectivity, vehicular connectivity, wireless ad hoc and mesh networks, could be potentially alternative solutions to facilitate NAN communications. However, the advantages and challenges of each of these enabling technologies are beyond the scope of this chapter. We are interested in exploring how the cyber interface of the NAN, referred to as the NAN GW, is able to monitor how much power is being distributed to a particular neighborhood by the corresponding Control Center at the Distribution Substation.

3.2 Building Area Network – BAN

Every building hooked up with the smart power grid maintains its own BAN. A BAN covers a number of apartments having HANs. The BAN smart meter, also known as the BAN GW, is typically set up at the building's power feeder. The BAN GW can be leveraged to monitor the energy demand and power consumption of the residents of the corresponding building. In order to facilitate BAN-HANs communication, cellular communication may be harnessed to cover more areas.

3.3 Home Area Network – HAN

A HAN is a subsystem within the smart grid dedicated to effectively manage the on-demand power requirements of the customers. For instance, *HAN*1 in Fig. 2 connects the home appliances (e.g., television, washing machine, oven, and so forth) in the customer's apartment to a HAN GW, which, in turn, communicates with *BAN*1. Indeed, the HAN GWs are the entities which facilitate M2M communications in the smart grid framework. In other words, the HAN GW of a residence communicates with the electrical appliances of that residence using the M2M communication paradigm. Next, we investigate the available M2M network protocols and compare their performances in the smart grid system.

4 Towards Effective Smart Grid Communication

In the early part of this section, we investigate the unique communication requirements of the smart grid followed by reviewing a number of available network technologies to facilitate M2M communication by fulfilling these requirements. Then, we present the best possible technology to be adopted for the M2M communication in the smart grid, and also show how simple yet effective enhancement can be made to the existing M2M technology to increase the effectiveness of the smart grid communication system.

4.1 Smart Grid Communication Requirements

The smart grid communication depends on two important parameters [26]. The first parameter is the communication delay while the second one is the large volume of messages. Actually these parameters can be relevant to the current big data concept, which is influenced by the velocity and volume requirements. Indeed, smart grid is projected to be part of the big data generating domains and these requirements should be carefully considered. If the Control Center at the NAN site of the smart grid misses any input from a HAN GW, this may influence the decision made by the Control Center. If any congestion happens at the BAN GW, the communication packets (referred to as the message) may be delayed to be sent to the NAN GW and the Control Center. To make matters worse, the message may also be dropped if the memory of the HAN GW becomes full either because of the limited processing power of the HAN GWs or many messages arriving from different M2M devices at the same time. In such a case, the HAN GW may request the M2M device to retransmit the required packets. This also increases the communication delay. While the overall smart grid communication delay in the order of a few milliseconds may not be practically achievable in large scale smart grid systems [26], focus should be given to minimizing the communication delay starting from the HAN level.

The work in [26] also indicated that the smart grid communication network should be able to accommodate more messages simultaneously without significantly influencing the communication delay. The massive amount of messages exchanged in the smart grid network will affect the available bandwidth. Therefore, it is important to take into consideration if it is possible to reduce the number of messages received from a large number of M2M devices at each HAN GW. This can assure that the total number of messages generated in the whole building does not disrupt the normal operation of the BAN GW.

4.2 M2M Network Technologies for HAN

To meet the demands of the smart grid communications, a number of short and medium range wireless technologies may be adopted. As mentioned earlier, the home appliances connected to a HAN form the M2M devices. To select an appropriate M2M network protocol in the smart grid HAN, we need to take into account the features of M2M devices in terms of low power consumption. Several low power and low cost technologies have emerged in the literature as enablers of M2M communication in the smart grid such as Bluetooth, IEEE 802.11 (WiFi), Ultra Wide Band (UWB), IEEE 802.15.4 ZigBee, 6LoWPAN, and so forth. The main technologies enabling HAN communications are delineated next.

4.2.1 IEEE 802.15.3a – Ultra-Wide Band (UWB)

UWB communication evolved for a number of applications that can be categorized into two types. The first type of application is for high data rate communications (typically over 1 Mbps) like High Definition Television (HDTV) transmission. The other type of applications with data rate below 1 Mbit/s (e.g., sensor networks) can also use UWB technology for exchanging information. The M2M devices in a HAN which can be considered as sensors may use UWB technology. The high power need of UWB, however, poses a significant challenge. In fact, this is one of the main reasons why the IEEE 802.15.3a task group was dissolved. Hence, further support for UWB may not be possible in the future if UWB is adopted as the communication technology in the smart grid HAN.

4.2.2 IEEE 802.11 – WiFi

The IEEE 802.11 protocol, popularly known as WiFi, is suited to facilitate high data rate applications over larger areas compared to UWB communications. For residential users, WiFi is the most recognized and commonly used technology. WiFi brings multiple band communication to take into account both coverage and capacity for residences and support IPv6 addressing. Despite these advantages, the main shortcoming of the WiFi technology is similar to that of UWB, in terms of the high power requirement of WiFi devices. As a consequence, WiFi is considered not to be practical for the smart grid M2M communications in the HAN level.

4.2.3 IEEE 802.15.1 – Bluetooth

The Bluetooth protocol has emerged as a popular choice for wireless connectivity for voice, data, and audio applications over short range. Because the Bluetooth protocol stack supports IP addressing, it can be adopted in the HAN communication. Unlike UWB and WiFi technologies, the Bluetooth protocol works well for low power/low data rate applications in peer-to-peer communications over a short distance. However, Bluetooth networks or "piconets" are able to support up to a limited number of devices simultaneously. To provide scalability among the M2M devices using Bluetooth, a HAN, therefore, requires to have a number of piconets (each having no more than 8 M2M devices). Each piconet consists of a master M2M node. The piconets can communicate with each other via the respective master nodes. However, this topology results in an increased communication delay. Another limitation of the Bluetooth technology is its periodical wake up and synchronization with the master node of the piconet. A Bluetooth device may take approximately 3 seconds to wake before the synchronization operation.

4.2.4 IEEE 802.15.4 – ZigBee

IEEE 802.15.4 ZigBee is now a well-known technology which is used in many home networking systems including HANs of the smart gid. ZigBee was developed, specifically, for the wireless devices requiring low power and long life time. The ZigBee network layer permits the M2M devices to form a cluster tree, self-healing mesh network, or star topologies. Thus, the HAN GW and the M2M devices within the HAN can be configured in a flexible manner. Also, a ZigBee device may consume only few milliseconds to wake up from the sleep state. This outperforms the wake up time performance of the Bluetooth or WiFi devices. Furthermore, a ZigBee device using the Carrier Sense Multiple Access With Collision Avoidance (CSMA/CA) protocol does not need to schedule special wake up events to communicate and maintain synchronization with the HAN GW. Thus, the ZigBee technology can perform and scale well in the smart grid HAN in contrast with the other competing technologies like UWB, WiFi, and Bluetooth.

In Fig. 3 [27], the power consumption comparison for each candidate technology (to be adopted as the enabler for the M2M communication in the Home Area Network of the smart grid) is shown. The comparison demonstrates that the Bluetooth and ZigBee protocols consume less transmission and reception power in contrast with those needed by WiFi and UWB technologies. In addition, IEEE 802.15.4 ZigBee with the Smart Energy Profile (SEP) Version 1.5 has more functionality compared to Bluetooth in the HAN communication context. Therefore, the Zigbee radio technology is adopted as the communication protocol in HANs owing to its low power demands (1–100 mW), simple network configuration and management features, and a reasonable communication range of 10–100 m.

While we expect the smart power grids to continue to evolve in the next decade, it is necessary to consider the most robust and reliable technology to facilitate M2M

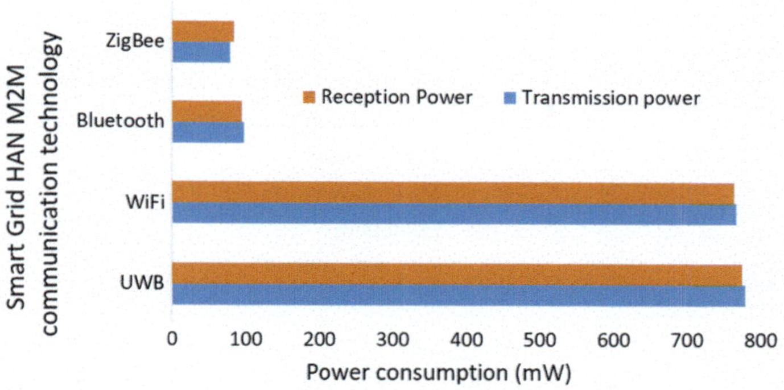

Fig. 3 Comparison of power consumption for different communication technologies that can be adopted for the M2M communication in the HAN of the smart grid

communication in the home area networks. In this article, we highlighted the infrastructure of a smart grid and described the major technologies available today for enabling the smart grid home area communication. We emphasized on choosing ZigBee protocol for the enabler of M2M communication in smart grid environments as it performs far better than other communication technologies such as UWB, WiFi, and Bluetooth. It should be emphasized, however, that the described M2M communication in the smart grid takes place within the considered home area network, and that the communications between the other entities (i.e., between home and building area networks, and between building and neighborhood area networks) are for data forwarding only. In other words, M2M communication is not occurring in the later entities in the presented smart grid model.

5 Security Challenge in Smart Grid Communication

As discussed in the earlier section, smart grid communication depends on two important parameters, namely the communication delay and large volume of messages in the smart grid. These two parameters lead to unique security issues for the smart grid communication. For instance, if the Control Center of the smart grid misses any input from a HAN smart meter, this may affect the decision made by the Control Center which may be critical in terms of demand response, load balancing, or even dynamic billing. Table 1 lists the power requirements of different home appliances that form the M2M nodes in a typical HAN. The smart grid communication system needs to be able to handle the message delivery to the Control Center via the BAN and NAN GWs with the minimum possible delay to ensure that near real-time decision making can be carried out regarding demand response and other operations. The power requirements of the HAN devices listed in Table I are sent to the respective BAN using the meter periodic data read. The size of each raw periodic request message is 32 bytes [3]. With the mandatory headers, the packet size

Table 1 Power demands of various home appliances which form the M2M nodes of a typical HAN

Electrical appliance (M2M node)	Power requirement (kW/hour)
Air conditioner	1
Refrigerator	0.2
Microwave oven	0.1
Light bulb	0.05
Personal computer	0.2

can be approximately (50 + 32=) 82 bytes. Additionally, there are TCP/IP headers and optional security headers if any security protocol is employed. If congestion happens at the BAN GW, the packet delivery to the NAN GW and the Control Center may be delayed. Furthermore, the packet may also be dropped if the Random Access Memory (RAM) and the on-chip flash of the BAN GW are occupied due to (*i*) multiple messages arriving from different HANs at the same time, and (*ii*) limited processing capability of the BAN GWs. In case of dropped packets, the BAN GW can request the HAN GW to retransmit the required packets. This also results in an increased communication delay. In practice, the smart grid communication latency should be in the order of a few milliseconds [3, 8]. However, this is difficult to achieve in large scale smart grid systems and security operations can lead to cryptographic overheads which may further increase the communication delay. As a result, how to keep the communication delay within reasonable margin while incorporating security and privacy operations has appeared as an important research focus in the smart grid research domain.

Furthermore, the work in [26] suggested that the smart grid communication network should be able to accommodate more messages simultaneously without any major influence on the communication delay. The large volume of messages in the smart grid communication will impact the required bandwidth in a significant way. For example, let us consider a model where a Control Center, connected to 10,000 building feeders/BAN GWs, serves 100,000 customers. Suppose that each HAN GW generates a message per second to the BAN GW in a typically power-intensive period, e.g., during a hot summer day when many customers simultaneously run their air-conditioners. The total number of generated messages per second is 100,000. The BAN GWs also generate messages to each other and also to the Control Center through the NAN GW. If we consider an average packet size of 100 bytes, the required transmission line bandwidth should be at least 800 Mbps.

From this simple example, it can be understood that the cryptographic operations to incorporate security and privacy to the information exchanged over the smart grid communication framework must be lightweight. Otherwise they will put more burden on the communication bandwidth and also lead to increased communication delay and communication overhead. Furthermore, the security headers should also be as small as possible so that they do not contribute to increased packet size. Hence, it may be inferred that a lightweight mechanism is critical for designing an effective authentication protocol for HAN/BAN/NAN GWs in the smart grid.

However, many of the traditional smart grid security protocols lack the detailed documentation, including the choice of adequate cryptosystems. Also, there is a lack of a holistic approach to use the same authentication procedure across all the communication networks of the smart grid. In other words, a particular authentication scheme can be effective in the HAN while it may incur significant communication overhead when applied to BAN/NAN. In a much needed holistic model, the BAN GW should authenticate the requesting HAN GWs while the NAN GW should be able to authenticate its BAN GWs in a lightweight yet secure manner. In addition, note that the cryptographic operations also contribute to significant computation cost, both in the receiver-end, which verifies the message, and in the Control Center's side. In the previous example containing 100,000 customers, if a smart meter sends message every second, the number of messages which need to be verified per second by the NAN GW is significantly high. Also, the processing delay at the respective smart meters for decrypting the incoming encrypted messages is substantially high. This affects the communication delay. Because the conventional Public Key Infrastructure (PKI) schemes are not adequate for the stringent time requirement of smart grid communications, a lightweight verification algorithm tailored for smart grid communications is required so that the incoming messages can be processed faster.

5.1 Envisioned Architecture of Smart Grid Communication and Security

First, we present our envisioned architecture of smart grid communication and security as depicted in Fig. 4. Here, the communication and security planes are split, and the interactions between them are shown using the blue arrows. For interested readers, the individual modules of the two planes can be found in [3]. Note that the blue arrows are bi-directional, which reflect the information exchanged between the two planes. For instance, the AMI captures some data from the smart meters and passes to the light-weight authentication protocol, which carries out some cryptographic functions and shares with the AMI. Our adopted light-weight authentication scheme, based on Diffie Hellman key exchanged protocol, (highlighted in red, in the figure) is discussed next.

5.2 Adopted Lightweight Authentication Scheme

Before delving into the adopted lightweight authentication Scheme [3] for the smart grid M2M communication, let us describe its core design objectives as follows.

1. Source authentication and message integrity: The smart meters need to be able to verify the origin and integrity of a received packet. For instance, if a BAN GW receives a packet from one of its HAN GWs, the BAN GW needs to authenticate

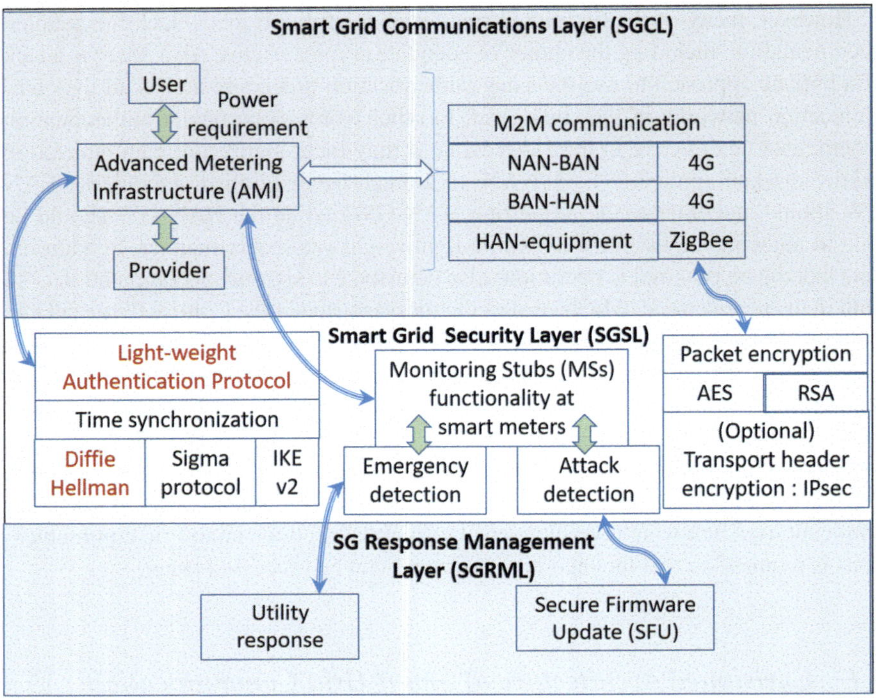

Fig. 4 Envisioned smart grid communication and security layers

the HAN GW. After successful authentication, it needs to check whether the packet is unmodified.

2. Low communication overhead and fast verification: The security scheme should be efficient in terms of small communication overhead and reasonable processing delay. Thus, a large number of message signatures from many smart meters should be verified in a short time interval.
3. Conditional privacy preservation: The actual identity of a smart meter (containing the name of the owner, the apartment number, and so on) should be protected/masked by adequate encryption technology.
4. Prevention of internal attack and upholding privacy: A HAN GW owner, having his/her own keying material, should not be able to intercept the neighboring HAN GWs' keying material. Thus, even if a smart meter is compromised, an adversary is unable to abuse the compromised smart meter to access and exploit other smart meters' important information.
5. Maintaining forward secrecy: A session key obtained from a set of long-term public and private keys cannot be compromised if one of the long-term private keys is compromised in the future.

Based on the above design objectives, a Diffie-Hellman key exchange based authentication procedure is adopted for facilitating light-weight, secure communication between the HAN/BAN/NAN GWs [3], as depicted in Fig. 5.

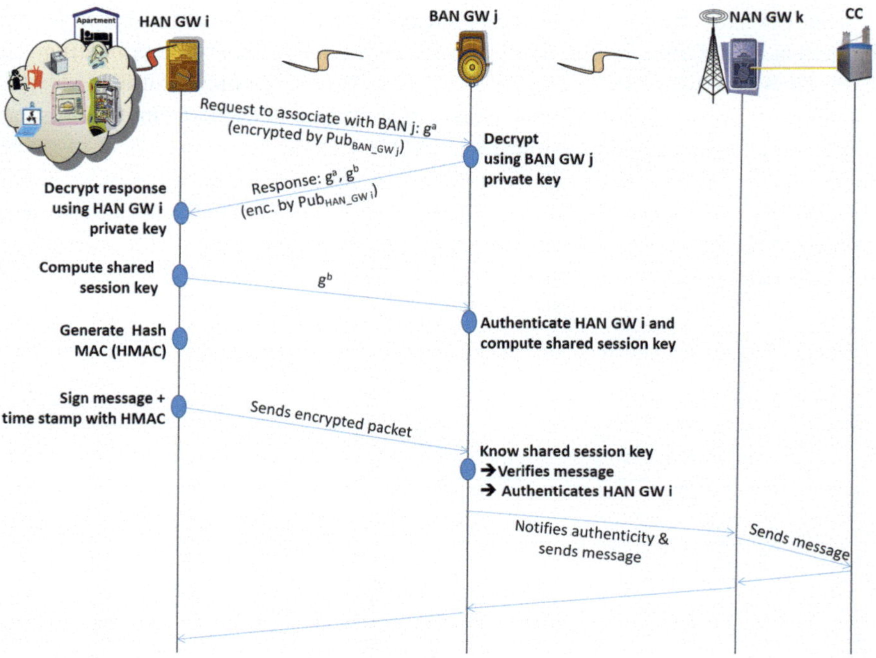

Fig. 5 Adopted lightweight message authentication scheme

Assume that HAN GW *i* and BAN GW *j* have their private and public key pairs, which have been distributed securely beforehand. This assumption is important, however, can be guaranteed a number of techniques for establishing secure channels in the literature. HAN GW *i* first chooses a random number *a*, and requests to be associated to BAN GW *j* by sending g^a. The request is encrypted using public key of BAN GW *j*. Here, $(g) = G$ is a group of large prime order *g* such that the Computational Diffie-Hellman assumption holds. In other words, if g^a and g^b are given for unknown integers *a* and *b*, computation of $g^{ab} \in G$ is hard. BAN GW *j* decrypts it using its private key and sends an encrypted response consisting of g^b, where *b* is a random number by using HAN GW *i*'s public key. After receiving BAN GW j's response packet, HAN GW *i* recovers g^a, g^b with its private key. If the recovered g^a is correct, BAN GW *j* is authenticated by HAN GW *i*. Then, with g^b and a, HAN GW *i* can compute the shared session key based on a function of g^{ab}.

The HAN GW *i* then sends g^b in the plaintext form. If the received g^b is correct, BAN GW *j* authenticates HAN GW *i* and is able to compute the same shared session key which the HAN GW *i* holds. To ensure data integrity in the subsequent transmissions, a Hash-based Message Authentication Code (HMAC) generation algorithm is used by employing the shared session key. The HMAC is based on the message and recorded time instance for sending the message to thwart replay attacks.

Figures 6, 7, 8, 9 and 10 demonstrate the performance of the adopted lightweight authentication method. In Fig. 6, the average communication overhead experienced at the BAN GW for an increasing number of HAN GWs from 10 to 140 is plotted. As shown in the figure, the adopted approach significantly outperforms the existing ellipict curve based authentication (ECDSA-256) for the growing number of smart meters or HAN GWs. On the other hand, the average delay experienced by the BAN GW for the same number of HAN GWs is shown in Fig. 7. The result in this figure shows that the adopted method incurs a much lower average delay as the authentication steps take shorter time compared to ECDSA-256 algorithm.

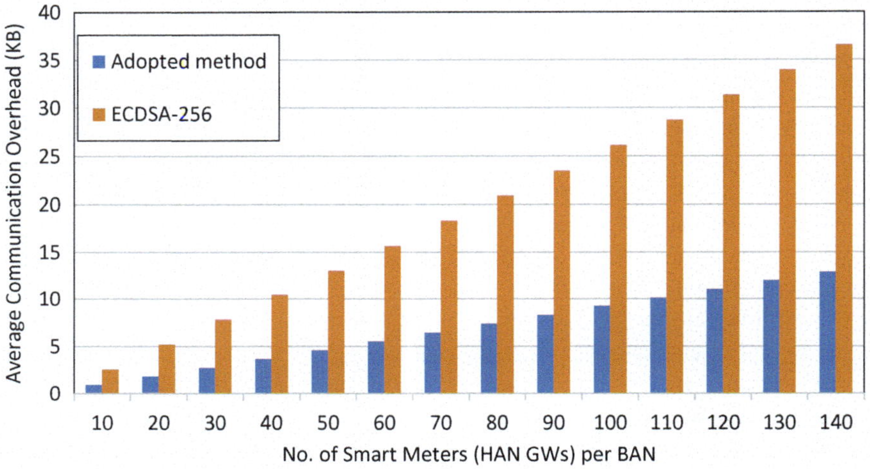

Fig. 6 Average communication overhead incurred at the BAN GW for a growing number of HAN GW

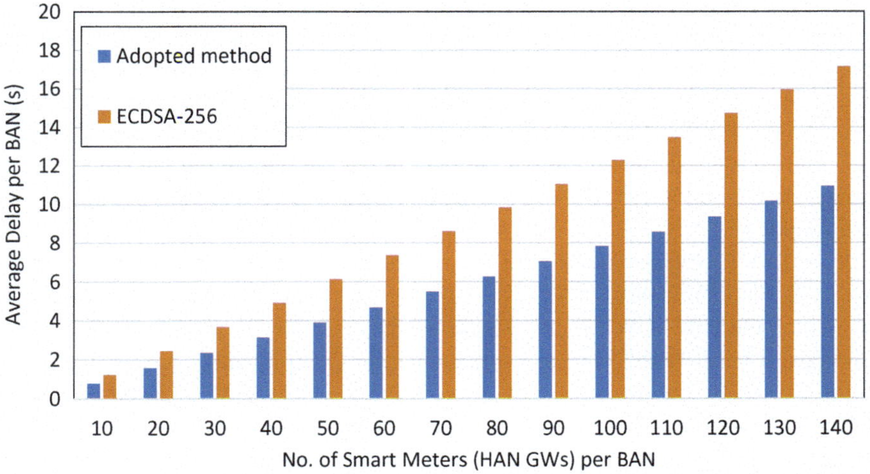

Fig. 7 Average delay incurred at the BAN GW for a growing number of HAN GWs

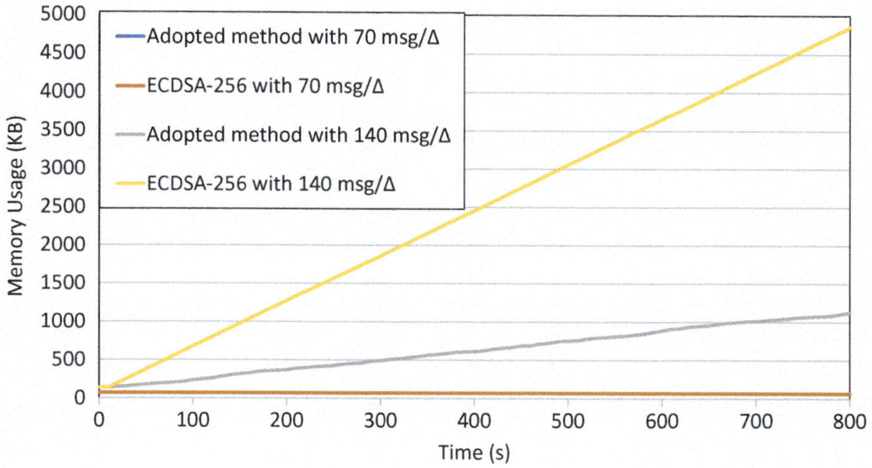

Fig. 8 Memory consumption comparison between the adopted and conventional ECDSA authentication algorithms for various message sizes received at the BAN GW

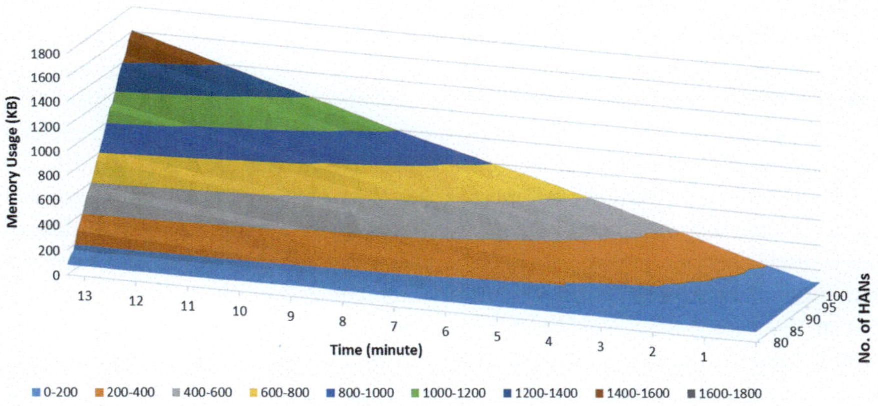

Fig. 9 Number of HANs supported by the existing method

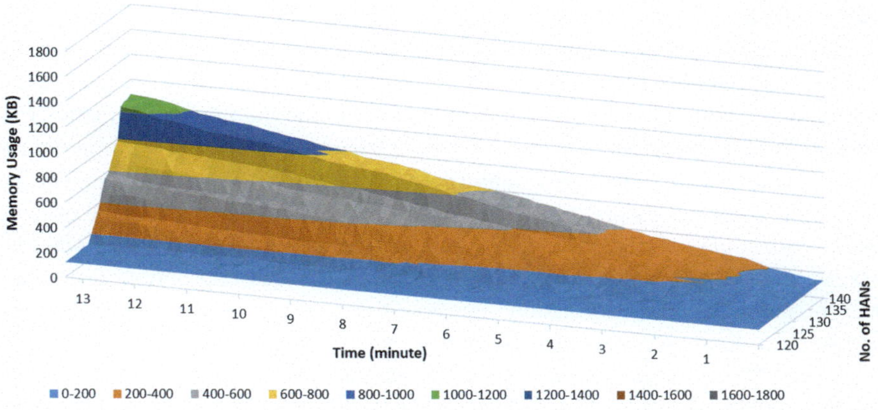

Fig. 10 Number of HANs supported by the adopted method

Next, the memory consumption at the BAN GW is compared for the adopted and existing methods in Fig. 8. The adopted method exhibits the same performance as that of the existing method for the packet generation rate of 70 messages/second. On the other hand, when the generation rate is doubled to 140 messages/second, the existing method exceeds the available memory (RAM plus flash memory) of the BAN GW while our adopted method still performs well within the available memory bound.

Figures 9 and 10 demonstrate the number of HANs supported by the existing and adopted methods, respectively. The results show that our adopted lightweight authentication method allows about 40% more HANs to be supported compared to the existing method.

6 Special Scenario: Securing Targeted Broadcast in Smart Grid

To improve the transmission delay and communication overhead issue discussed earlier in the chapter, it is essential to minimize message transmission as much as possible in the smart grid while securely and flexibly authenticating the M2M devices and smart meters. In this section, we consider a special scenario whereby the smart grid Control Center needs to transmit different messages to different neighborhoods and different customers. The Control Center has two options, either to encrypt and dispatch these messages sequentially in a unicast manner, or broadcast a single encrypted message and allow the appropriate receivers (neighborhood/building/home users) to extract the information intended for them. With the broadcast methodology, the Control Center benefits by only sending once, saving precious processing, memory, and network resources and also time. This is because this method allows the receiver or user-side smart meters to decrypt and access the information intended only for them. The problem, however, is how to facilitate this in the smart grid communication?

To better understand the problem, consider the following scenario with two small towns called Port Hope and Wellington in Ontario, Canada. In the scenario, the Control Center may only want to send sensitive information to the Port Hope residences. Is it possible for the Control Center to flexibly only send one encrypted information so that all residences at Port Hope can read the information while the users of Wellington are not able to decrypt it? Consider another scenario in which the customers of the smart grid subscribe to different packages, and are due to receive different energy pricing. Then, if the Control Center wants to share pricing events with some specific package holders, is it possible for the Control Center to flexibly send just a single encrypted message so that the appropriate users can decrypt and access the information? For both the scenarios, how to issue a secure targeted broadcast message opens up an interesting problem.

To solve the aforementioned problem, in the remainder of this chapter, we adopt the Key Policy Attribute Based Encryption (KP-ABE) [24] to offer a flexible

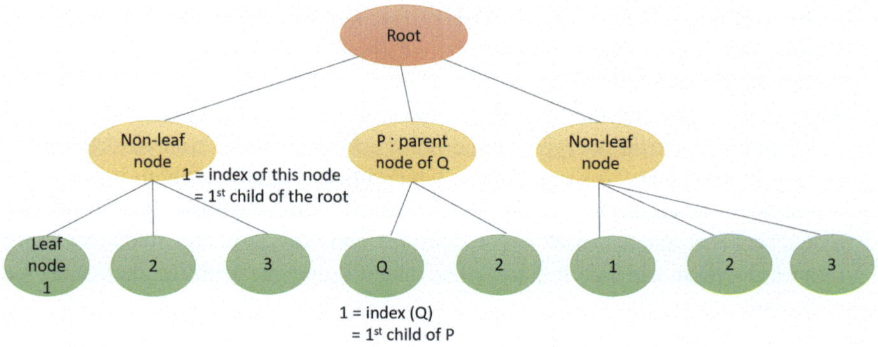

Fig. 11 A sample tree-based access structure (**A**) employing KP-ABE in the considered smart grid

solution [2]. Why do we choose this methodology for this purpose? Because, KP-ABE is able to provide an efficient public key cryptography primitive for one-to-many encryption, which fits well with the targeted broadcast scenario in the smart grid context.

Let us first describe the KP-ABE preliminaries. Suppose that all the smart grid users have some attributes. Choosing a proper set of attributes is the initial step to construct an efficient KP-ABE targeted broadcast for smart grid.

Assume that **U** is the universal set of all user attributes. Suppose that **A** is a tree-based access structure comprising simple AND, OR logic gates. A sample tree-based access structure for the considered KP-ABE targeted broadcast for the considered smart grid is depicted in Fig. 11. Every non-leaf node denotes a logic gate, and has a threshold. If its threshold is equal to one, it is an OR gate. On the other hand, if its threshold equals its children number, it is expressed as an AND gate. In addition, each leaf node is considered as an attribute. All the nodes in the access tree are ordered by index numbers as demonstrated in the figure.

The access tree, **A**, is served as input to the KP-ABE algorithm, which executes in the following steps.

1. In the initialization step, KP-ABE uses a cyclic group G1 generated by a generator g of prime order p and another cyclic group G2 of the same order to construct a bilinear map represented by e: G1 × G1 ➔ G2. Then, the Control Center of the smart grid generates the KP-ABE public key set PK = (T1, T2, …, TN, Y) where $Y = e(g,g)^y$ with a random number $y \in Z_p^*$, and master key set MK = (t1, t2, …, tN, y).
2. Then, the message to be broadcast from the smart grid's Control Center, denoted by M, is encrypted under k attributes with a random number input s, and the ciphertext C is generated.
3. The secret decryption key, D, is then generated by taking as inputs the leaf nodes of the access tree A (i.e., the attributes) and master key MK. The secret decryption key of node x, denoted by D_x can be expressed as: $g^{\frac{q_{parent(x)}(index(x))}{t_i}}$.

4. Each smart meter obtains its KP-ABE private key from the Control Center. This is a one-time key-distribution process, which should be performed over secure transmission like Secure Socket Layer (SSL).
5. When there is an encrypted broadcast transmission from the Control Center, the smart meters perform decryption operation on the received ciphertext by using its decryption key D. This is a recursive decryption operation, which starts on the root node of the access tree A, and then recursively iterates through all non-leave nodes, finally to the leaf nodes. The complex conditions are satisfied if and only if the ciphertext satisfies the tree, and the original message M can, thus, be recovered.

Thus, the ciphertexts are linked with the smart grid users' different attributes, while the user secret keys are defined with the access structures on the attributes. Hence, a user is able to decrypt the ciphertext only if the ciphertext attributes satisfy the user's access structure.

Now, an example is presented to illustrate how KP-ABE targeted broadcast performs secure broadcast to targeted users. Suppose that there are three types of messages exchanged between the Control Center and the smart meters: maintenance schedule announcement, real-time price event, and meter firmware update request. The attributes of the smart meters to correspond to these three message types are: location (zip-code/address), subscription package profile, and firmware version. Let P1, P2, and P3 represent governmental, industrial, and residential subscriptions, respectively. If the Control Center needs to send the real-time price event only applicable to the industries, it should encrypt the message using the proposed KP-ABE targeted broadcast with the industrial subscription attribute. Similarly, it can also be performed to announce for maintenance periods to residents of a specific location. For example, the Control Center can broadcast a firmware update request targeting residential subscribers, which can be decrypted only by the residential smart meters. This type of flexible targeted broadcast is secure and efficient, where the encrypted messages can be decrypted by the targeted group alone without the need for multiple encrypted message generation accompanied with multiple unicast transmissions.

In Table 2, an efficiency comparison among the secure broadcast using KP-ABE and that with two other implementations of CP-ABE (namely BCP-ABE1 and BCP-ABE2) are listed. The size of ciphertext, private key, and public key are compared amongst these three schemes. Here, n and r denote the total number of users and number of revoked users, respectively. t denotes the access structure size, which can be at most l. The number of attributes linked with the private key in the considered ABE schemes is represented by k, which can be maximum up to m. Table 2 demonstrates that KP-ABE has smaller cipher text size compared to both the BCP-

Table 2 Comparison of KP-ABE and other ABE-based broadcast methods

Scheme	Cipher size	Private key size	Public key size
KP-ABE broadcast	$(k + 1) + 1$	$2t$	$(m + 4) + (2n-2)$
BCP-ABE1	$(t + 1) + 1$	$k + 2$	$m + l + 3 + (2n-1)$
BCP-ABE2	$(t + 1) + 2r$	$(k + 2) + 2$	$(m + l + 3) + 4$

ABE1 and BCP-ABE2. BCP-ABE1 performs better than the KP- ABE approach in terms of the private key size. But, the public key size for KP-ABE is much smaller than both BCP-ABE1 and BCP-ABE2 methods. With this trade-off, KP-ABE outperforms the other ABE methodologies and emerges as a viable candidate to be used for secure targeted broadcast in the smart grid communication.

7 Concluding Remarks

The smart grid communication is often taken for granted due to simplified application or overlaying of the communication infrastructure with the power grid system. Researchers and practitioners of the smart grid have traditionally ignored the implications brought by the merging of the physical and cyber components. By incorporating communication framework on top of the energy distribution system, cyber threats like Distribute Denial of Service (DDoS) attacks, worm propagation, Man-in-the-Middle Attacks, and many other threats can jeopardise the security and privacy of the grid operator as well as the users. This becomes more complicated as much of the communication takes place as among IoT sensors, machines, and other things deployed all over the smart grid including home area networks and energy distribution sites. Such M2M communication has its own requirements leading to difficulty in incorporating legacy authentication methods borrowed from the existing literature. As a solution, we demonstrated how to adopt a lightweight authentication scheme for the various hierarchical networks in the smart grid. In addition, a specific scenario where targeted broadcast is required in the smart grid was discussed and a unique authentication methodology based on Key Policy Attributed Based Encryption (KP-ABE) was presented.

References

1. Fadlullah ZM, Quan DM, Kato N, Stojmenovic I (2014) GTES: an optimized game-theoretic demand-side management scheme for smart grid. IEEE Syst J 8(2):588–597
2. Fadlullah Z, Kato N, Lu R, Shen X, Nozaki Y (2012) Toward secure targeted broadcast in smart grid. IEEE Commun Mag 50(5):150–156
3. Fouda MM, Fadlullah ZM, Kato N, Lu R, Shen X (2011) A lightweight message authentication scheme for smart grid communications. IEEE Trans Smart Grid 2(4):675–685
4. Fadlullah Z, Fouda M, Kato N, Shen X, Nozaki Y (2011) An early warning system against malicious activities for smart grid communications. IEEE Netw 25(5):50–55
5. Fadlullah ZM, Fouda MM, Kato N, Takeuchi A, Iwasaki N, Nozaki Y (2011) Toward intelligent machine-to-machine communications in smart grid. IEEE Commun Mag 49(4):60–65
6. Wei C, Fadlullah ZM, Kato N, Stojmenovic I (2014) A novel distributed algorithm for power loss minimizing in smart grid. In: 2014 IEEE international conference on smart grid communications (SmartGridComm). IEEE
7. Fadlullah ZM, Duong MQ, Kato N, Stojmenovic I (2013) A novel game-based demand side management scheme for smart grid. In: 2013 IEEE wireless communications and networking conference (WCNC). IEEE

8. Fouda MM, Fadlullah ZM, Kato N, Takeuchi A, Nozaki Y (2012) A novel demand control policy for improving quality of power usage in smart grid. In: 2012 IEEE global communications conference (GLOBECOM). IEEE
9. Fadlullah ZM, Nozaki Y, Takeuchi A, Kato N (2011) A survey of game theoretic approaches in smart grid. In: 2011 International conference on wireless communications and signal processing (WCSP). IEEE
10. Fouda MM, Fadlullah ZM, Kato N (2010) Assessing attack threat against ZigBee-based home area network for smart grid communications. In: The 2010 international conference on computer engineering & systems. IEEE
11. Yuan Y, Li Z, Ren K (2011) Modeling load redistribution attacks in power systems. IEEE Trans Smart Grid 2(2):382–390
12. Goyal V, Pandey O, Sahai A, Waters B (2006) Attribute-based encryption for fine-grained access control of encrypted data. In: Proceedings of the 13th ACM conference on computer and communications security – CCS 06. ACM Press
13. IEEE (2009) Ieee, conference drive smart grids. [Online]. Available: http://www.eetimes.com/electronics-news/4081867/IEEE-conferencedrive-smart-grids
14. Alaa Y, ElAttar HM, Digham F, Afify LH, Elbadawy H (2017) Lte dynamic scheduling scheme for massive m2m and h2h communication. In: 2017 IEEE 8th annual ubiquitous computing, electronics and mobile communication conference (UEMCON). pp 478–482
15. Cha I, Shah Y, Schmidt A, Leicher A, Meyerstein M (2009) Trust in m2m communication. IEEE Veh Technol Mag 4(3):69–75
16. Chen D, Brown J, Khan JY (2014) Performance analysis of a distributed 6lowpan network for the smart grid applications. In: 2014 IEEE ninth international conference on Intelligent Sensors, Sensor Networks and Information Processing (ISSNIP), April 2014, pp 1–6
17. Starsinic M (2010) System architecture challenges in the home m2m network. In: 2010 IEEE Long Island systems. Applications and technology conference, IEEE
18. Ericsson GN (2010) Cyber security and power system communicationessential parts of a smart grid infrastructure. IEEE Transactions on Power Delivery 25(3):1501–1507
19. Metke AR, Ekl RL (2010) Smart grid security technology. In: 2010 Innovative smart grid technologies (ISGT). IEEE
20. Privacy-friendly aggregation for the smart-grid. http://research.microsoft.com/apps/pubs/?id=146092
21. Kgwadi M, Kunz T (2010) Securing RDS broadcast messages for smart grid applications. In: Proceedings of the 6th international wireless communications and mobile computing conference on ZZZ – IWCMC 10. ACM Press
22. New study stresses need for unied security approach for smart grid (2011) http://www.smart-meters.com/the-news/smart-grid-news/2389-new-study-stresses-need-for-unied-security-approach-for-smart-grid
23. Attrapadung N, Imai H (2009) Conjunctive broadcast and attribute-based encryption. In: Pairing-based cryptography pairing 2009. Springer, Berlin Heidelberg, pp 248–265
24. Bethencourt J, Sahai A, Waters B (2007) Ciphertext-policy attribute-based encryption. In: 2007 IEEE symposium on security and privacy (SP 07). IEEE
25. Ruj S, Nayak A, Stojmenovic I (2011) A security architecture for data aggregation and access control in smart grids. CoRR, vol. abs/1111.2619, 2011. [Online]. Available: http://arxiv.org/abs/1111.2619
26. Hauser CH, Bakken DE, Dionysiou I, Gjermundrod KH, Irava VS, Helkey J, Bose A (2008) Security, trust, and QoS in next-generation control and communication for large power systems. International Journal of Critical Infrastructures 4(1/2):3
27. Lee J-S, Su Y-W, Shen C-C (2007) A comparative study of wireless protocols: Bluetooth, UWB, ZigBee, and wi-fi. In: IECON 2007 –33rd annual conference of the IEEE industrial electronics society. IEEE

Zubair Md. Fadlullah is currently an Associate Professor with the Computer Science Department, Lakehead University, and a Research Chair of the Thunder Bay Regional Health Research Institute (TBRHRI), Thunder Bay, Ontario, Canada. He was an Associate Professor at the Graduate School of Information Sciences (GSIS), Tohoku University, Japan, from 2017 to 2019. He also served at GSIS as an Assistant Professor from 2011 to 2017. His main research interests are in the areas of emerging communication systems like 5G New Radio and beyond, deep learning applications on solving computer science and communication system problems, UAV based systems, smart health technology, cyber security, game theory, smart grid, and emerging communication systems. He was a recipient of the prestigious Dean's and President's Awards from Tohoku University in March 2011, and the IEEE Asia Pacific Outstanding Researcher Award in 2015 and NEC Tokin Award for research in 2016, for his outstanding contributions. He has also received several best paper awards at conferences including IWCMC, Globecom, and IC-NIDC. He is a Senior Member of the Institute of Electrical and Electronics Engineers (IEEE), and IEEE Communications Society (ComSoc).

Mostafa M. Fouda received his Ph.D. in Applied Information Sciences in 2011 from Tohoku University, Japan. He received B.Sc. with honors in Electrical Engineering, and M.Sc. in Electrical Communications from Benha University, Egypt, in 2002 and 2007, respectively. He is currently serving as a Postdoctoral Researcher at Department of Electrical and Computer Engineering, Tennessee Tech University, TN, USA. Also, He has been serving at Faculty of Engineering at Shoubra, Benha University, Egypt, since 2002, and in June 2016, he was promoted to the position of an Associate Professor. He was a recipient of the prestigious 1st place award during his graduation from the Faculty of Engineering at Shoubra in 2002. His research interests include Cyber Security, Internet of things, 5G communications, software-defined networks (SDNs), and smart grid communications. He is a Senior Member of the Institute of Electrical and Electronics Engineers (IEEE), and IEEE Communications Society (ComSoc).

Combating Intrusions in Smart Grid: Practical Defense and Forecasting Approaches

Zubair Md. Fadlullah and Mostafa M. Fouda

Abstract The smart grid represents one the biggest growth potentials of the Internet of Things (IoT) use case. The smart grid communication is a typical example of the inter-machine communication, which is popularly referred to as the Machine to Machine (M2M) communications whereby the deployed "things" such as smart meters and numerous sensors require none/minimal human intervention to characterize power requirements and energy distribution. The plethora of sensors have the ability to report back critical information like power consumption of the users and other monitoring signals on power quality to the control center. Thus, the energy distribution grid is coupled with the IoT sensing and delivery networks in the smart grid. However, this inherent design of the smart grid poses a significant security challenge, particularly from the networking domain, in terms of malicious events like Distributed Denial of Service (DDoS) attacks against smart meters and other devices. In this chapter, we overview two attack scenarios in the smart grid, at the Home Area Network (HAN) and the Building Area Network (BAN), respectively. HAN is a key part of the smart grid communications framework through which the customers are able to communicate with the electricity provider. In a HAN, there is typically a smart-meter and a number of electric appliances which communicate over ZigBee (IEEE 802.15.4) wireless technology. Even though ZigBee incorporates some security features, the technology still suffers from a number of security vulnerabilities in the smart grid environment. To demonstrate this, we present a HANIdentifier (HANId) conflict attack against ZigBee for HAN communications and demonstrate the impact of the attack on the smart grid communications. Then, an appropriate framework is presented to prevent the attack from taking place. Next

Z. Md. Fadlullah (✉)
Computer Science Department, Lakehead University, Thunder Bay, ON, Canada

Thunder Bay Regional Health Research Institute, Thunder Bay, ON, Canada
e-mail: zubair.fadlullah@lakeheadu.ca

M. M. Fouda
College of Engineering, Tennessee Tech University, Cookeville, TN, USA

Faculty of Engineering at Shoubra, Benha University, Benha, Egypt
e-mail: mfouda@ieee.org

Z. Md. Fadlullah, A.-S. Khan Pathan (eds.), *Combating Security Challenges in the Age of Big Data*, Advanced Sciences and Technologies for Security Applications, https://doi.org/10.1007/978-3-030-35642-2_10

in the chapter, we introduce more advanced concept using Gaussian Process to model the malicious attacks on a broader network level which may compromise the security and privacy of smart grid users. Based on our Gaussian Process based model, a lightweight and practical method to forecast intrusions in the smart grid communication network is proposed. By leveraging the proposed approach, the smart grid control center is able to predict malicious attacks so that early action can be taken to protect the smart grid from being adversely affected. Simulations results demonstrate the viability of the proposed forecasting method.

1 Introduction

Recently, the Internet of Things (IoT) and Machine-to-Machine (M2M) [1] communications have digitally transformed our society. Various use cases of IoT and M2M are starting to appear a plethora of domains including smart homes, smart cities, smart health, autonomous driving, smart grid, and so on [2–6]. In particular, the proliferation of IoT and the machine-centric communication among the IoT devices spurred the growth of the smart energy grid sector [7–11]. The smart grid is basically an energy grid comprising smart or intelligent capability to permit the power providers, distributors, and customers to be able to maintain a real-time or even near real-time awareness of their respective operating needs and capabilities. By leveraging this awareness, the smart grid is able to generate the electricity by appropriately matching the specified demands of the customers as well as distribute it to the consumers in an efficient fashion. Typically, there are numerous electrical equipment connected in an intricate way to the smart grid. The interconnection between the huge number of devices is to ensure that they have the ability to report to the smart grid's control center regarding the power demand, consumption, and quality related information. This allows the power distribution network to make "smart" decisions by anticipating how much power will be needed to homes and factories and providing them with the demanded power at a significantly lower cost by diverting energy from distributed and renewable sources like solar and wind generators. While there has been a lot of effort given toward designing an integrated communication and power grid system, serious attention is required for smart grid security [3]. Just to illustrate how security is important at a typical home, consider a single residence scenario whereby an end-user may have many electric appliances. Those appliances need to be connected to the home network's coordinator, with which they may exchange information on power requirements and usage by utilizing IEEE 802.15.4 or ZigBee specification (zigbee). ZigBee is a wireless and media access protocol for cheap and power-saving personal area networking devices. The smart grid, along with the rest of the IoT community, started using protocols like ZigBee particularly in HANs. Although the 802.15.4 specification supports a number of security features using a link-layer security mechanism, those security features are not designed to address some fundamental yet significantly impactful attacks. In this chapter, we investigate such a fundamental link-layer attack that exploits the HANIdentifier (HANId) conflict message and demonstrate how it degrades the smart grid communication performance in the HAN level.

Furthermore, because smart grid communication is based on the conventional networking technologies, the same security concerns which frequently arise in contemporary networks are also relevant in the smart grid scenario. In particular, the smart meters are vulnerable to network threats. Even though the smart meters typically need to carry out authentication to communicate with the other smart meters and devices, the authentication scheme itself can be targeted by Distributed Denial of Service (DDoS) attackers. In this chapter, we consider the spread of worms in the smart grid that compromises a number of machines (i.e., smart meters and electrical appliances), which begin to transmit malicious authentication requests to the victim smart meters. These malicious cases observed in a smart grid network are reported to the upper network tiers. By this way, the information is conveyed to the smart grid's control center. By modeling the malicious attack event as a Gaussian process, the control center has the ability to forecast the occurrence of future malicious events in the smart grid system.

The cyber threats like DDoS attacks are likely to have more impact on smart grid communication due to the involvement of so many electrical equipment on the customer's premise. By adopting an early prediction technique for malicious threats on the smart grid, it may be possible to promptly respond to protect the cyber physical system from being overwhelmed or even compromised by the malicious entity. In order to design such a forecasting framework, however, we need to consider a lightweight algorithm and also the fact that human intervention is inappropriate for the machine type communication in the smart grid. The nature of the smart grid requires that the machines within the smart grid cyber physical system should have an adequate framework to forecast malicious and/or abnormal events such as cyber-attacks and equipment failures. The machines in the smart grid are deployed in different network settings, e.g., residential networks and wide area networks. The smart meters deployed in these different networks may be harnessed to create an information sharing network. This kind of information sharing can offer an important resource for developing global and timely assessments of emerging malicious threats against smart grid communication.

The remainder of the chapter is organized as follows. First, we present the literature review on link layer security vulnerability at the HAN level, followed by the research work conducted on early prediction of network threats in the smart grid context. Then, the considered architecture for smart grid communication is described. Then, a unique attack model based on the HANId conflict message is presented. Next, the need for forecasting malicious threats against the smart grid is discussed. Our solution adopting Gaussian process based attack prediction method is then presented and evaluated. Finally, concluding remarks on caveats and future directions are provided.

2 Background and Related Research

In order to meet all the functional requirements of a smart grid from a communication viewpoint, it is essential to take into consideration many standards. The Institute of Electrical and Electronics Engineers (IEEE)'s initiative to define these standards and provide guidelines on the smart grid functionality is worth following in this regard. The IEEE devised standards merged the recent advances in power engineering, communications, and Information Technology (IT) in a holistic manner. This led to the formation of the IEEE P2030 which consists of a number of taskforces focusing on the integration of various energy sources, load side requirements, cyber security, and so on (kowa). By this way, they committed to take into account different aspects of power engineering along with IT and communications techniques. It was also worth mentioning that the IT group would investigate issues such as privacy, security, data integrity, interfaces, and interoperability in the smart grid. However, the security has been traditionally dealt in a loose manner by not considering the unique needs of the smart grid cyber physical system. On the other hand, the communications technology group was handed the responsibility to find the communication requirements of the heterogeneous devices used in the smart grid. In other words, the aim of the power group was to define boundaries on power generation, transmission, and distribution by considering the customers. In the end, the policies designed by the aforementioned work-groups appeared to be rather broad in nature resulting in coarse design directives to enforce security in the smart grid communications.

One of the first notable research works in the smart grid security was [12], which came up with a computer network oriented security and authentication management system for servicing actions and commands request in the smart grid control center. The shortcoming of the work, however, consisted in its sole focus on only securing the electric power systems and electric circuits in the host/control center area. In other words, the work failed to address the holistic smart grid security need by missing the smart grid communication framework in its model. Indeed, the work in [13] demonstrated how crucial it is to take into account the power system communication and address the cyber security elements. This particular work advised researchers to address and resolve a plethora of cyber security vulnerabilities affecting the smart grid communications. For instance, integrated SCADA/EMS systems and administrative office IT environments were shown to potentially result in significant security threats. The work also demonstrated the security vulnerability emanating from the adoption of broadband to connect smart meters with the central control system of the smart grid. While the utility companies are interested in transferring data to the residences which may include price information and special offers, such data may also contain control signals, which may raise delicate security issues and privacy breaches.

Next, in [14], it was revealed that the smart grid deployments have to meet strict security demands. For example, strong authentication is regarded to be a paramount requirement for all customer devices and appliances connected to the smart grid network. As a consequence, the work stressed upon the need of a scalable key and trust management system, tailored to the particular requirements of the utility company, given the large number of customer devices which can be impacted in the advent of a security breach like a malicious intrusion or cyber attack.

The research work mentioned above, however, do not address the security requirements of the IEEE 802.15.4 (ZigBee) technology, which is leveraged for the Home Area Network (HAN) communications. In this chapter, by providing a broad smart grid communications framework, we point out a security concern in the ZigBee-oriented home area network and try to deal with this security vulnerability [15].

On the other hand, regression has been used in many areas of computer science and engineering for predicting various things in a plethora of models. In particular, in the applied network security field, a number of machine learning based security mechanisms for predicting anomalies and clustering abnormal behaviors from normal baselines have been carried out. The smart grid domain can also be benefited by the already mature science of prediction. However, the prediction tasks in the smart grid are often relevant to load forecasting, storage battery depletion warning, and so forth. For instance, in [16], accurate real-time load forecasting was reported to be essential for reliable as well as efficient operation of a smart power grid system. The work in [16] utilizes the accurate reporting of the emerging Advanced Metering Infrastructure (AMI) to track the incoming load requests from Plug-in Hybrid Electric Vehicles (PHEVs). It exhibits the benefits of the smart use of AMI data in generation planning and load forecasting. Also, a data forecasting scheme for estimating the electricity consumption beforehand was proposed in the work in [17]. This method adopts three-point Gaussian quadrature approach to build the forecasting model. Another approach for predicting power use was developed in [18] that leveraged kernel regression based on local models for large-scale data mining scenarios in smart grid. The work in [19] surveyed various techniques for load forecasting in power grids that consist of various techniques such as regression, exponential smoothing, iterative re-weighted least-squares, stochastic time series, and so forth. The survey in [19] also stresses on an important trend, i.e., hybrid mechanisms are required to forecast events in the power grid that combine two or more of these techniques.

However, the afore-mentioned research endeavors only take into account the smart grid power consumption information and do not investigate whether such models can be effectively employed for forecasting abnormal events in the smart grid power distribution and/or communication networks. Although Gaussian processes have been considered in many learning scenarios in literature [20], they have not been utilized in learning abnormal modes of operation in smart grid. In this chapter, we simplify the modeling of Gaussian process in smart grid communication for predicting malicious events, which may disrupt the operation of the smart meters.

3 Smart Grid Communication Framework

This section covers the basics of smart grid communication framework. The smart grid power transmission and distribution system delivers power, which is generated from the power plant, to the customers over a transmission substation and a number of distribution substations. The transmission substation delivers power from the power plant through high voltage transmission lines (usually over 230 kilo volts) to distribution substations. The distribution substations are placed in different regions, and are responsible for converting the electric power into medium voltage levels and distributing this medium-voltage level power to the building-feeders. In order to make it usable by the consumers, the building feeders have to convert the medium voltage level into a lower level.

Our considered smart grid communication system, i.e., the cyber element, is separated from the physical component (power transmission and distribution system), and can be regarded as an information sharing network consisting of a number of hierarchical components. For communication, however, the above consideration may not be hold because the communication links have different requirements than those of the power lines. The transmission substation and the control centers of the distribution substations are connected with one another in a meshed network, which can be constructed over optical fiber technology. The remaining components of the considered smart communication topology is divided into a number of networks, as depicted in Fig. 1, which feature real-life set-ups of a city or a metropolitan area. Generally speaking, a city can be divided into a number of regions (e.g., wards), each of which is serviced by a distribution substation. Every region consists of several neighborhoods, each neighborhood has many buildings and houses. Each building may have a number of apartments and each house/apartment of a building may have a number of rooms. Our smart grid communication framework is inspired from this real-life planning of a metropolitan area as follows.

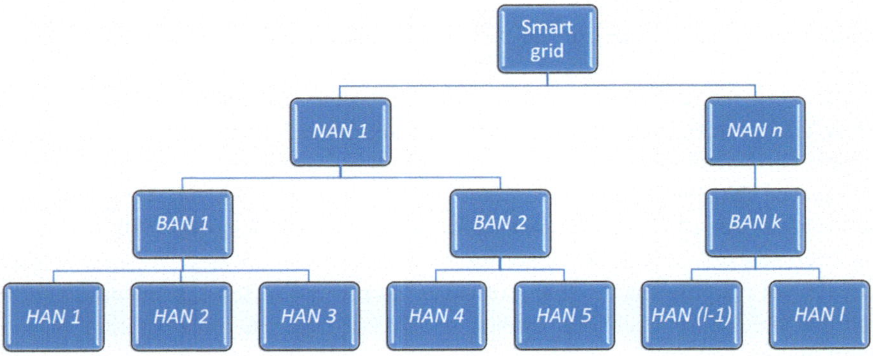

Fig. 1 Considered smart grid architecture. NAN, BAN, and HAN refer to Neighborhood Area Network, Building Area Network, and Home Area Network, respectively

The communication architecture for the lower distribution network (beneath the control center) is divided into a number of hierarchical networks, namely Neighborhood Area Network (NAN), Building Area Network (BAN), and Home Area Network (HAN). This design is, in spirit, similar to the smart grid system model and processes described in the work in [21]. Each NAN can be regarded to be composed of a number of BANs. On the other hand, every BAN is composed of a number of apartments. The apartments have their respective local area networks, each of which is referred to as a HAN. Additionally, there are advanced meters, referred to as the smart meters, which are deployed in the smart grid architecture that represent AMI for enabling an automated, two-way communication between the utility meter and the utility provider. The smart meters are equipped with two interfaces: *(a)* power reading interface and (*b*) communication gateway interface. The smart meters are cyber physical devices and have both physical and cyber interfaces. Its physical interface deals with the electricity supply and delivery system while the cyber element acts as a communication gateway (GW). Therefore, for brevity, the smart meters used in the NAN, BAN, and HAN are referred to as NAN GW, BAN GW, and HAN GW, respectively. Furthermore, it is also worth mentioning that based on the existing standards of smart grid, Internet Protocol (IP)-based communication is preferred to allow virtually effortless inter- connections with HANs, BANs and NANs.

3.1 Neighborhood Area Network – NAN

A NAN is a localized network of the considered smart grid communication topology. A NAN consists of one or more cellular base stations and a number of BANs. Notice that the cellular framework adopted for smart grid communications should be different from the existing ones used for providing other services, e.g., Internet. This assumption is made in order to prevent network congestion. Also, it is possible to avoid security threats arising from the Internet that may have impact on the delay-sensitive smart grid communications. The NAN GW can monitor how much power is being distributed to a particular neighborhood by the corresponding control center at the distribution substation.

3.2 Building Area Network – BAN

Every building connected to the smart power grid maintains its own BAN. A BAN consists of a number of apartments having HANs. The BAN smart meter/GW is typically set up at the building's power feeder. The BAN GW can be used to participate in power demand and response negotiation. Similarly, a BAN GW needs to be authenticated with the NAN GW.

However, the smart meter's authentication scheme may be susceptible to Distributed Denial of Services or DDoS attacks. For instance, we consider a worm propagation scenario in the smart grid whereby the worm forces the infected host, e.g., a smart meter or some consumer-device, to inject bogus or mal-formed authentication packets to legitimate smart meters. The smart meters at the HAN or BAN are, thus, forced to perform expensive signature verifications to authenticate the compromised hosts. As a consequence, the victim smart meters end up exhausting their already limited memory and processing resources.

In the next section, we propose an early warning system for smart grid to forecast the afore-mentioned DDoS attack.

3.3 Home Area Network – HAN

A HAN is a subsystem within the smart grid dedicated to effectively manage the on-demand power requirements of the end-users. For example, *HAN* 1 in Fig. 1 is responsible for the equipment (such as television, washing machine, oven, and so forth) in the first apartment of the considered building to a HAN GW, which, in turn, communicates with *BAN* 1. It is worth noting that a HAN can also consist of renewable and/or backup power sources including electrical vehicle, solar panel, battery storage, small wind turbine, and so on.

4 HAN ZigBee Attack Model

In this section, we address a type of attack towards the ZigBee based wireless network at the HAN. The attack exploits the HANIdentifier (HANId) conflict. In a HAN of an apartment employing ZigBee, there are a smart meter acting as the HAN coordinator and a group of nodes, which represent the home appliances. Let us refer to the smart meter's unique identifier as the HANId. The members of a given HAN know their HANId. In case there exists more than one HAN coordinator, which runs in the same operating space, a HANId conflict may occur. We derive this attack model in spirit with the one for wireless sensor networks based on the IEEE 802.15.4 technology. In the event of such a HANId conflict, the HAN coordinator can detect the conflict through its received beacons or one of the home appliances belonging to the HAN could notify the HAN coordinator on receiving signal from two HAN coordinators with the same HANId. Upon receiving the notification, the HAN coordinator performs the conflict resolution procedure (mac). This mechanism mainly covers the channel scans and coordinator realignment procedure that includes choosing a new HANId and broadcasting it to all its HAN nodes. After resynchronization with beacons, the network is ready to communicate in a stable way. Thus, the conflict resolution is resolved. We present an attack scenario whereby a malicious user, using a compromised home appliance, can frequently send forged conflict notification messages to the HAN coordinator and force the coordinator to carry out the conflict resolution procedure repeatedly. A smart attacker, which is able to easily produce HANId conflict notification messages by setting the related field in the message frames, is then able to exploit these forged messages to prevent or significantly delay communication between the smart meter (i.e., the HAN coordinator) and the home appliances. We demonstrate the impact of the HANId conflict attacks on smart grid communications through computer simulations in MATLAB. Particularly, we simulate the considered HANId conflict attack to study its influence on HAN communications delay. The simulation model is typically a HAN employing IEEE 802.15.4 (ZigBee) technology. The simulated HAN consists of a HAN coordinator (i.e., smart meter) and 15 devices that are connected to the HAN coordinator in a star topology formation. Several of the devices act as malicious users (i.e., attackers) in the course of the simulation. After the association process between the HAN coordinator and the devices, the attack-node(s) is/are assumed to transmit fake HANId conflict notification messages at random intervals. When the HAN coordinator receives a conflict notification, it performs the appropriate handling mechanism as part of the IEEE 802.15.4 specification [22].

An important parameter in the simulation model is the interval time between receiving a conflict event at the HAN coordinator and the end of the realignment process. We set this parameter to 3 s as being observed in [23]. Because the HAN coordinator is not able to process any other conflict notifications during this realignment process, it just ignores any HANId conflict notification during the stated 3 s periods. The simulation time is set to 100 s within which the attacker(s) are assumed to send 10 fake HANId conflict notification messages at arbitrarily chosen time intervals. Because the attack-node(s) may not be synchronized with one another, some attacker(s) may transmit the fake conflict during the realignment process of the coordinator. This may lead to missing or ignoring some conflict notifications from the attacker(s).

The HAN coordinator parameters are referred to as Tl, T2, and T3. Tl is defined as the maximum number of conflicts for an attacker while T2 is the maximum number of HANId conflicts in a duration time (T3) for an attacker. Three scenarios are taken into account with the number of attacker(s) varying from one to four. The results are depicted in Figs. 2, 3, and 4.

First, we consider a scenario whereby only Tl is considered as shown in Fig. 2. Two cases are chosen, namely for Tl = 3 and Tl = 4. In the first case, when the maximum number of allowed attacks is set to three (i.e., T1 = 3), the total conflict resolution delay at the HAN coordinator continues to increase. For example, for a single attacker model with T1 = 3, the detection latency is 5 s in contrast with almost 20s of detection delay when there are four attackers in the system. On the other hand, when the system is relaxed to permit one more HANId conflict (i.e., when T1 = 4), the detection delay approaches approximately 30s for four attackers. This shows that even with conventional threshold-based detection schemes, multiple attackers may have a significant impact on the smart grid communications for nearly 20–30s, during which other legitimate devices are deprived of the utility service as they are detached from the HAN coordinator, which goes through the realignment process.

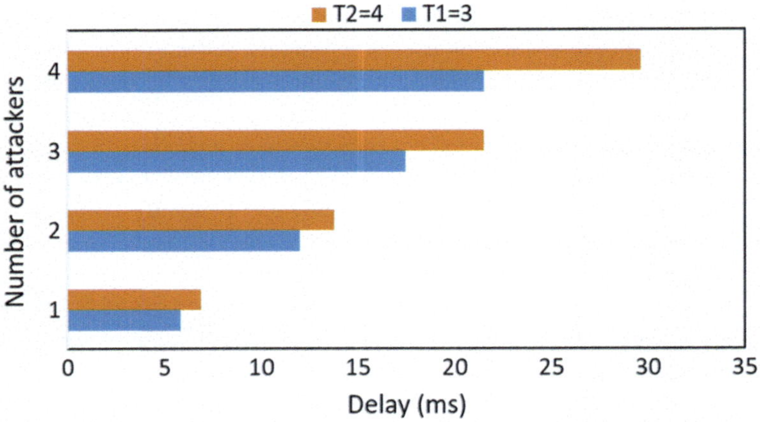

Fig. 2 Decision delay when using only a single threshold (T1) for up to 4 attackers

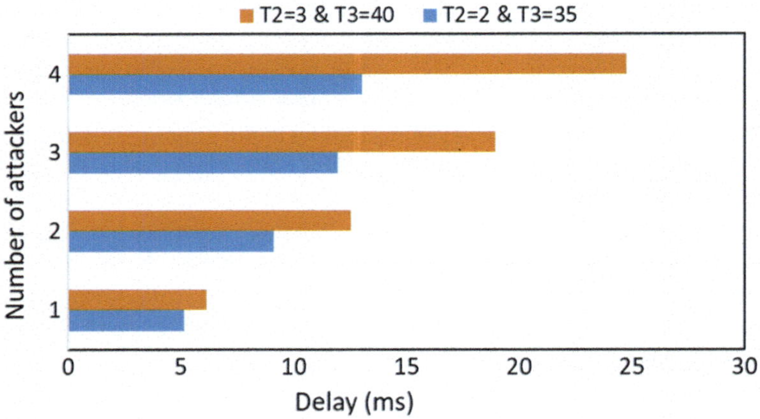

Fig. 3 Attack resolution delay with two thresholds (*T*2 and *T*3) for number of attackers varied from 1 to 4

In the second scenario, only T2 and T3 are taken into account and the total conflict resolution delays are plotted for varying numbers of HANId conflict attackers as depicted in Fig. 3. When T2 = 2 and T3 = 35 s, the HAN coordinator takes about 5 s and 13 s to detect the cases comprising a single attacker and 4 attackers, respectively. On the other hand, when T2 is set to 3 for T3 = 40s, the HAN coordinator experiences a significantly longer time (approximately 25 s) to resolve the HANId conflict notifications from the 4 attackers. The reason for this is due to the fact that the conflict resolution mechanism is executed by the HAN coordinator repeatedly in this case.

In the third and final scenario, we merge the influence of all the parameters (i.e., T1, T2, and T3) at the HAN coordinator to investigate the influence of the attack

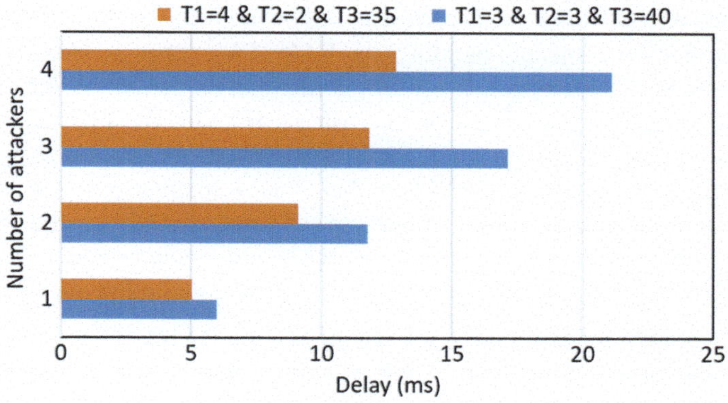

Fig. 4 Impact of using all three thresholds (T1, T2, and T3) on the HANId attack resolution delay

resolution on smart grid communications. As evident by the results shown in Fig. 4, the HANId conflict resolution delays are significant for the different parameter settings and increase more with the growing number of attackers.

Thus, these results elucidate that the HANId conflict attacks affect the smart grid communications if they are simply ignored. Therefore, it is essential that we prevent them from occurring in the first place by envisioning an appropriate smart grid communication framework.

5 HanID Attack Prevention Framework

In this section, we present a smart grid communications framework so that the HANId conflict attack is prevented from taking place. Our proposed framework is assumed to consist of two data repositories at the NAN GW and the BAN GW, respectively. The NAN GW repository contains the building information and the BANIds that refer to unique identifiers to represent each BAN in the considered neighborhood area. When a new building is constructed in a neighborhood, the new BAN GW dispatches a request, over cellular link, to its corresponding NAN GW manager to register with the NAN GW. NAN GW constructs a BANId for this new building area network by incrementing the total number of already existing buildings in its covered neighborhood by one. It is worth noting that the NAN GW may use other information related to the building (e.g., building name, owner name, and so on) to construct the BANId. Then, it shares the registered BANId with the appropriate BAN GW. The BAN GW saves this information in its own repository. Thus, the NAN GW is able to also track the number of buildings in an efficient manner. On the other hand, at the BAN GW, the BANId is used to construct the HANIds of all the HANs belonging to that particular BAN. For a particular apartment's HAN, the HANId consists of its BANId as the preamble following by the apartment number. In this simple model, the apartment number is used to follow the preamble to obtain a unique HANId for each apartment. Following the activation of the HAN of a new apartment in a specific building, the BAN GW creates a HANId and other details related to the apartment. It is worth noting that the BAN GW may use other information regarding the apartment to create its HANId. By this way, in such a managed framework, it is not possible for an apartment to obtain a duplicate HANId. This eliminates the chance of HANId conflict attacks. In other words, if a compromised device connected to a HAN attempts to carry out a HANId conflict attack, the HAN GW will immediately know that this HAN is the only one to subscribe that particular HANId. At this stage, the HAN GW can take the following actions:

(1) The HAN GW considers the node, which sent the HANId conflict message, as malicious, and ignores future HANId conflict messages from the malicious node. Note that the HAN GW does not entirely block the malicious node as this may deny service to the appliance and isolate it from the power supply.
(2) The HAN GW then downloads and forces the secure framework update for the node acting in a malicious manner.
(3) It informs the owner (e.g., by sending a text message or email to his/her cellphone) regarding the event so that he/she may manually check and repair the equipment.

Consider the worst case scenario whereby there are two adjacent buildings. In the absence of our envisioned framework, the HANs in the two neighboring buildings may be assigned the same HANId. Since the appliances in these two HANs are to operate in a work space quite close to each other, they would continuously send HANId conflict messages to their respective HAN GWs. Our adopted solution deals with this scenario by assigning unique BANIds to each building networks which are, in turn, used to formulate unique HANIds even in this exceptional case.

6 DDoS Attack Model in Smart Grid Network

In this section, we present a DDoS attack model against smart grid communication framework which is based on the DDoS targeting broadcast authentication in wireless sensor networks. In the considered attack model, an attacker is assumed to be able to eavesdrop, inject, and modify packets transmitted in the smart grid network. Additionally, the attacker is assumed to possess access to at least a smart meter (e.g., a HAN GW) to infect the network by running computationally resourceful nodes, e.g., laptops and workstations. Furthermore, the attacker may also employ multiple smart meters to run distributed attacks concurrently. Particularly, through worm infection attacks, the attacker may exploit already compromised multiple colluding smart meters in different hierarchical networks of the smart grid. However, we assume that the infected smart meters cannot compromise the cryptographic secrets of other smart meters that are used during authentication.

Authentication is a basic yet critical security service in smart grid networks. The authentication primarily involves the smart meters in the different component networks of the smart grid. For instance, a HAN GW needs to be authenticated with its corresponding BAN GW prior to communication.

7 Light-Weight Forecasting of Malicious Events in Smart Grid

In this section, we initially discuss a set of guidelines for designing an effective early warning system for smart grid. The discussion reveals why Gaussian process regression is chosen for formulating our adopted prediction scheme.

7.1 Smart Grid Early Warning System Design Guidelines

The main objective of an early warning system is to reliably and accurately forecast problems in the smart grid communication network and raise alerts regarding the problems. Since the smart grid users are paying customers, they expect to get notified quickly about problems emerging in the grid. Particularly, the problems that lead to service interruption or service denial should be detected as early as possible. If such events can be predicted and notified to the users ahead of time, it is an even better option. For the smart grid control center, early forecasting and warning can help to promptly localize network problems so that they may be addressed more quickly in order to provide uninterrupted service to the customers.

Traditionally the problem detection/prediction is limited to the control center of the existing energy grids. It is worth noting that a smart grid network problem can be anywhere, including HANs, BANs, and NANs. The emergence of the problem may be related to either communication or power- related issues, or both. In other words, a wide variety of problems such as network congestion, power distribution anomalies (e.g., voltage level spikes), malicious attacks, and so on may fall into the scope of prediction. Another design consideration of the prediction system should be whether it should be centralized or distributed. Even though distributed prediction systems may be preferred to centralized ones, we should ask ourselves whether it is practically feasible. Individual smart meters at home or building premises have limited processing and memory resources. As a result, it may not be practical to integrate the early forecasting feature on these devices. On the other hand, the smart grid control center or the NAN GWs may be equipped with forecasting capability because of more advanced resources. For instance, unusual activities monitored at a building level can be reported to the NAN, and then forwarded to the control center. The control center can, then, predict whether a problem is imminent at the respective smart meter, contact with the smart meter confronting the anomaly or malicious intrusion so that it may take appropriate action to mitigate the problem. Additionally, the control center can issue emergency notifications to the building or neighborhood smart meters, or even other regional control centers, to inform them regarding possible occurrence of similar abnormal activities.

Next, an effective early warning system should not be limited to dealing with only a single type of malicious activity. For brevity, in this chapter, only a particular malicious use case involving the DDoS attack model is discussed. From the design perspective, the prediction scheme should be common to different malicious activities in order to avoid additional complexity because of the adoption or combination of different methodologies. A common architecture for smart grid warning system should be non-parametric. This means that the warning system should not be influenced by the different types of inputs or patterns obtained from various malicious activities. For instance, Gaussian process based techniques are well known in developing various spatial and temporal models since Gaussian processes offer a principled, probabilistic approach to facilitate machine learning [24]. In the remainder of the chapter, we propose the adoption of Gaussian process regression to forecast the malicious events in the smart grid. Gaussian process formulation can also be applied in a similar way to other malicious activities such as equipment malfunction, worm propagation, and so forth.

7.2 Gaussian process formulation in smart grid

Let us consider the concept of random variables to represent abnormal and/or malicious activity features in smart grid communication. These random variables may be the number of defective smart meters in a building, the fraction of malicious authentication attempts in a given unit of time, and so on. By considering such a collection of random variables in the smart grid communication, it is possible to formulate a Gaussian process. In the assortment of random variables, any finite subset of these variables can be found to have a joint multi-variate Gaussian distribution. It is worth noting that a Gaussian distribution is fully specified by a mean vector and a covariance matrix. On the other hand, a Gaussian process is fully characterized by a mean function and a covariance or kernel function. Also, it is worth reminding that the valid covariance functions result in positive semi-definite covariance matrices. For two arbitrary inputs to the covariance function, the corresponding functional outputs indicate the level of similarity.

Gaussian processes can be leveraged to offer a rich class of models and when fitted appropriately, they are quite flexible. Gaussian Process Regression is an example of such flexible features of the Gaussian processes that we adopt for our forecasting purpose in the smart grid.

7.3 Gaussian Process Regression

Gaussian Process Regression can be defined as a Bayesian data modeling method, which accounts for uncertainty in a comprehensive manner. Similar to other Bayesian-based inference methods, a Gaussian process consists of a prior and a posterior. The distributions are defined over functions using the Gaussian process that is used as a prior for the Bayesian inference. This prior can be flexibly derived from the training or observation data. This means that we obtain prior beliefs regarding the form of the underlying model. Through observations or experiments, data regarding the model are obtained. For example, in the smart grid environment, the data gathered from the smart meters can be utilized to form the prior beliefs of a Gaussian process which represents the different aspects of the smart grid communication.

Let us suppose that the prior belief about the considered function conforms to a Gaussian process with a prior mean and covariance matrix. Through the Gaussian Process Regression, samples of the function at different locations in the domain are observed. When a set of observation points and their corresponding real valued observations are known, it is possible to estimate the posterior distribution of a new point. It is worth noting that this posterior distribution is also Gaussian with mean and variance functions. The optimal parameters of the Gaussian process are computed by maximizing the log likelihood of the training data with respect to the parameters. Thus, it is possible to make predictions for unseen test cases by estimating the posterior.

7.4 Covariance function selection

Next, we discuss the importance of covariance function selection in order to appropriately model the Gaussian process. This is because it must generate a non-negative definite covariance matrix for any set of points or observations in the smart grid. While stationary and non-stationary covariance functions can be used, the selection of a covariance function relies on the specific nature of the targeted problem. In the aforementioned forecasting approach based on Gaussian Process Regression, we adopt a composite covariance function since it is more flexible in the considered smart grid scenario that sums covariance contributions from long and short term trends, the periodic component, and fluctuations comprising various observation lengths. Two isotropic squared exponential covariance functions are employed to characterize the long and short term trends. The periodic component is the product of a smooth periodic covariance function and another isotropic squared exponential covariance function without the latent scale. The fluctuations are denoted by an isotropic rational quadratic covariance function.

8 Performance Evaluation of the Forecasting Scheme

In order to demonstrate the performance of the proposed forecasting scheme, we describe the following scenario comprising the considered DDoS attack against the smart grid environment. In the experimental scenario, we suppose a large building comprising 20 apartments, i.e., the BAN GW is assigned to 20 HAN GWs. The BAN GW is assumed to have ten times higher specification/configuration than that of a HAN GW. The BAN GW is considered to be a smart meter with 160 MHz CPU, 128 KB RAM, and 1 MB flash memory. Also, let us assume that half of these apartment-owners did not install the latest patches/updates on their smart meters. This results in half of the HAN GWs to get infected by a worm, which gradually propagates over the considered building area network. The remaining apartment owners are assumed to have the latest updates on their smart meters and therefore, they are susceptible to the worm infection. The HAN GWs, infected by the worm, gradually commence generating malicious authentication requests to the BAN GW at variable rates. The BAN GW sends responses to their authentication requests. The authentication process is simulated with Elliptic Curve Digital Signature Algorithm (ECDSA) [2]. Thus, the malicious smart meters attempt at consuming the rather constrained memory resources available at the BAN GW to deny the legitimate authentication requests from the HAN GWs which are not infected.

First, we verify the impact of this simple yet effective attack against the BAN GW through experiments conducted in MATLAB. Fig. 5 shows the memory consumption at the victim BAN GW over time for various average rates of malicious authentication requests, denoted by n. In the conducted simulations, the value of n is varied from 100–200 per attack launch interval, Δ. The value of Δ is set to a particularly large value of 10 s. Also, it is worth noting that n is contributed by the 10 infected HAN GWs during Δ. The results in Fig. 5 show that the BAN GW memory

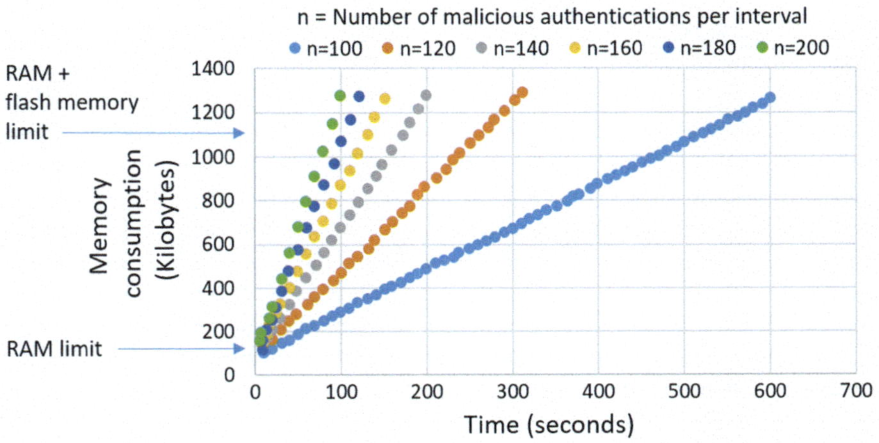

Fig. 5 Memory consumption at the victim smart meter at the BAN level for different attack rates in terms of number of malicious authentications per time interval ranging from 100 to 200

is fully consumed, i.e., the BAN GW is overwhelmed, more rapidly as the value of n grows. For example, for the lowest considered DDoS attack rate (with $n = 100$), it takes 500 s to exceed the BAN GW memory limit. On the other hand, for the higher attack rates with $n = 180$ and $n = 200$, it takes just about 80–100 s to consume the entire memory of the BAN GW. Then, the BAN GW remains busy to verify the malicious authentication signatures, and its limited yet precious memory is not available to service legitimate requests from the remaining uninfected HAN GWs. As a result, the legitimate smart meters are denied communication with the BAN GW and are not able to specify their power requirements to the smart grid control center.

Next, we report the simulation result of our forecasting algorithm for $n = 100$ as illustrated in Fig. 6. In this case, a training time of 400 s is considered to clearly explain how the results are obtained in the conducted experiment. The training and test data sets are highlighted in Fig. 6. Notice that the orange lines show the probabilistic predictions, in terms of the projected highest and lowest values of memory usage due to the attack. Also, the average of the highest and lowest values of the predicted data corresponds well with the actual test data set. Given this information conveyed from the BAN GW to the NAN GW, the smart grid control center evaluates the forecasting time, and alerts other BAN GWs about possible malicious threats for adequate response such as requesting their respective HAN GWs to update the firmware to prevent worm infection. Furthermore, it is worth reminding that for a substantially large training time, the predicted attack occurrence time corresponds with the test data with significant accuracy as depicted in Fig. 6.

In Fig. 7, the time instances at which BAN GW memory is predicted to become exhausted, for various values of n, are plotted. Note that the delay of BAN GW to control center communication has not been considered in this result since it does not

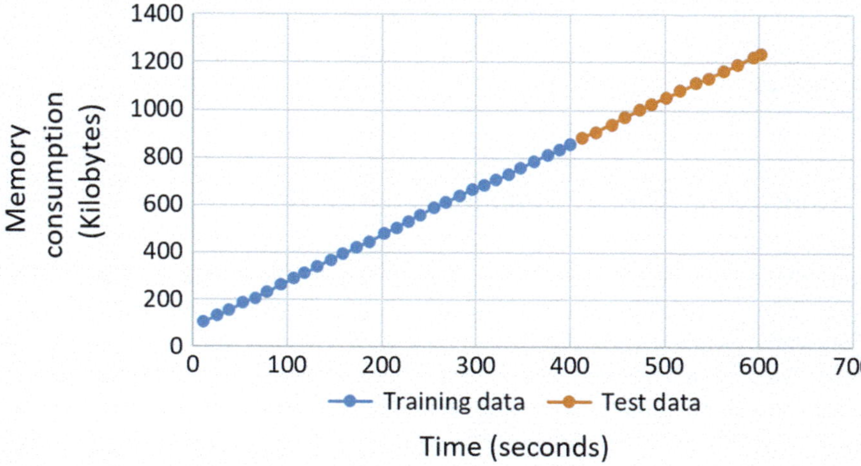

Fig. 6 Adopted forecasting method to predict when the victim BAN device will be overwhelmed by the DDoS attack

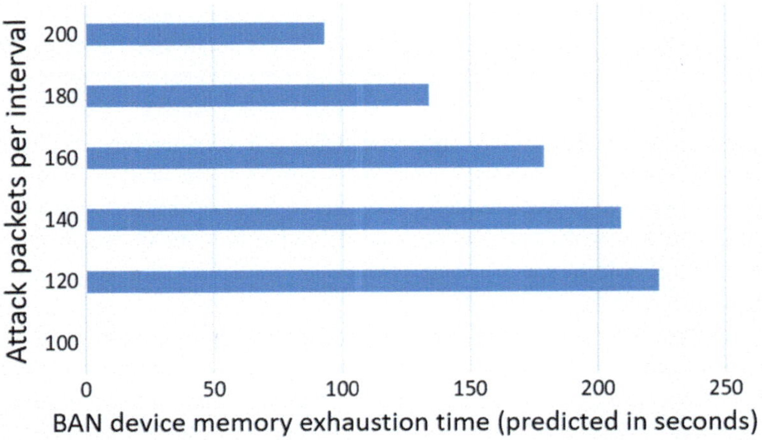

Fig. 7 The time predicted by the proposal to the BAN device memory exhaustion. The training time of 50 s is considered for different DDoS attack rates from 100 to 200 malicious authentication packets per time interval

change the purpose and the fundamental results of the conducted simulations. The training time considered in Fig. 7 is much lower than that considered in the result shown in Fig. 6. The reason behind this is the fact that higher values of n cause the BAN GW to be overwhelmed by the corresponding DDoS attack fairly quickly. Therefore, the training time is set to 50 s, i.e., lower than the actual value of the BAN GW memory exhaustion time (i.e., 80 s) for the considered highest DDoS attack rate (i.e., n=200). As evident in Fig. 7, the time required by the control center to predict the attack occurrence is reasonable. However, the only limitation is that the BAN GW memory exhaustion due to the lowest attack rate is not predicted by the control center because it does not manifest enough attack features during the short period. Hence, it is also important to adopt different windows of training time simultaneously in order to trap the effect of both the moderate and high rates of DDoS attacks.

9 Concluding Remarks and Future Directions

In this chapter, we described a HANId conflict attack and showed how to address the attack by designing an efficient framework for smart grid communication. Then, we presented a Gaussian Process Regression based model for forecasting malicious attacks, which may arise in emerging smart power grids. The framework employs probabilistic distribution to forecast if some abnormal mode of operation may disrupt smart grid communications. Simulation results revealed that the proposed method can warn the smart grid users regarding the malicious DDoS attacks beforehand. In addition to the described attack in this chapter, other malicious threats and anomalies (e.g., abnormal voltage surges and fluctuations, equipment failure, and so forth) can also be predicted using the proposed method so that the control center can instruct smart meters to take relevant actions against such anomalies in a prompt and efficient manner. However, for smart grid cyber physical systems with background network traffic, we need to establish a baseline for assessing errors to differentiate actual abnormal activities. Machine learning techniques like deep neural networks can be used to address such challenges in the future.

References

1. Fadlullah ZM, Fouda MM, Kato N, Takeuchi A, Iwasaki N, Nozaki Y (2011b) Toward intelligent machine-to-machine communications in smart grid. IEEE Commun Mag 49(4):60–65
2. Fouda MM, Fadlullah ZM, Kato N, Takeuchi A, Nozaki Y (2012) A novel demand control policy for improving quality of power usage in smart grid. In: 2012 IEEE global communications conference (GLOBECOM). IEEE
3. Fouda MM, Fadlullah ZM, Kato N, Lu R, Shen X (2011) A lightweight message authentication scheme for smart grid communications. IEEE Trans Smart Grid 2(4):675–685
4. Wei C, Fadlullah ZM, Kato N, Stojmenovic I (2014) A novel distributed algorithm for power loss minimizing in smart grid. In: 2014 IEEE international conference on smart grid communications (SmartGridComm). IEEE
5. Fadlullah ZM, Quan DM, Kato N, Stojmenovic I (2014) GTES: an optimized game-theoretic demand-side management scheme for smart grid. IEEE Syst J 8(2):588–597
6. Fadlullah ZM, Nozaki Y, Takeuchi A, Kato N (2011c) A survey of game theoretic approaches in smart grid. In: 2011 international conference on wireless communications and signal processing (WCSP). IEEE, Piscataway
7. Fadlullah ZM, Duong MQ, Kato N, Stojmenovic I (2013) A novel game-based demand side management scheme for smart grid. In: 2013 IEEE wireless communications and networking conference (WCNC). IEEE, Piscataway
8. Fadlullah Z, Kato N, Lu R, Shen X, Nozaki Y (2012) Toward secure targeted broadcast in smart grid. IEEE Commun Mag 50(5):150–156
9. Fadlullah Z, Fouda M, Kato N, Shen X, Nozaki Y (2011d) An early warning system against malicious activities for smart grid communications. IEEE Netw 25(5):50–55
10. Drew G (2008) Zigbee wireless networking, 1st Edition, Newnes Books, Elsevier
11. Malak T. Al-Nory (2019) Optimal decision guidance for the electricity supply chain integration with renewable energy: aligning smart cities research with sustainable development goals. IEEE Access, vol. 7, pp. 74996–75006
12. Hamlyn A, Cheung H, Mander T, Wang L, Yang C, Cheung R (2007) Network security management and authentication of actions for smart grids operations. In: 2007 IEEE Canada electrical power conference. IEEE, Piscataway
13. Ericsson GN (jul 2010) Cyber security and power system communication – essential parts of a smart grid infrastructure. IEEE Trans Power Deliv 25(3):1501–1507
14. Metke AR, Ekl RL (2010) Smart grid security technology. In: 2010 innovative smart grid technologies (ISGT). IEEE, Piscataway
15. Fouda MM, Fadlullah ZM, Kato N (2010) Assessing attack threat against ZigBee-based home area network for smart grid communications. In: The 2010 international conference on computer engineering & systems. IEEE, Piscataway
16. Alizadeh M, Scaglione A, Wang Z (2010) On the impact of SmartGrid metering infrastructure on load forecasting. In: 2010 48th annual Allerton conference on communication, control, and computing (Allerton). IEEE, Piscataway

17. Wang XJ, Wang HJ, Hou LQ (2010) Electricity demand forecasting based on threepoint gaussian quadrature and its application in smart grid. In: 2010 international conference on computational intelligence and software engineering. IEEE, Piscataway
18. Kramer O, Satzger B, Laessig J (2010) Power prediction in smart grids with evolutionary local kernel regression. In: Lecture notes in computer science. Springer, Berlin/Heidelberg, pp 262–269
19. Alfares HK, Nazeeruddin M (2002) Electric load forecasting: literature survey and classification of methods. Int J Syst Sci 33(1):23–34
20. Alecu T, Voloshynovskiy S, Pun T (2006) The Gaussian transform of distributions: definition, computation and application. IEEE Trans Signal Process 54(8):2976–2985
21. Niyato D, Wang P, Han Z, Hossain E (2011) Impact of packet loss on power demand estimation and power supply cost in smart grid. In: 2011 IEEE wireless communications and networking conference. IEEE, Piscataway
22. 802.15.4 (2006): IEEE standard for information technology–Local and metropolitan area networks- Specific requirements- Part 15.4: Wireless Medium Access Control (MAC) and Physical Layer (PHY) Specifications for Low Rate Wireless Personal Area Networks (WPANs).
23. Ning P, Liu A, Du W (2008) Mitigating DoS attacks against broadcast authentication in wireless sensor networks. ACM Trans Sens Netw 4(1):1–35
24. Zhang J, Porras P, Ullrich J (2008) Gaussian process learning for cyber-attack early warning. In: Proceedings of the 2008 SIAM international conference on data mining. Society for Industrial and Applied Mathematics, Philadelphia

Zubair Md. Fadlullah is currently an Associate Professor with the Computer Science Department, Lakehead University, and a Research Chair of the Thunder Bay Regional Health Research Institute (TBRHRI), Thunder Bay, Ontario, Canada. He was an Associate Professor at the Graduate School of Information Sciences (GSIS), Tohoku University, Japan, from 2017 to 2019. He also served at GSIS as an Assistant Professor from 2011 to 2017. His main research interests are in the areas of emerging communication systems like 5G New Radio and beyond, deep learning applications on solving computer science and communication system problems, UAV based systems, smart health technology, cyber security, game theory, smart grid, and emerging communication systems. He was a recipient of the prestigious Dean's and President's Awards from Tohoku University in March 2011, and the IEEE Asia Pacific Outstanding Researcher Award in 2015 and NEC Tokin Award for research in 2016, for his outstanding contributions. He has also received several best paper awards at conferences including IWCMC, Globecom, and IC-NIDC. He is a Senior Member of the Institute of Electrical and Electronics Engineers (IEEE), and IEEE Communications Society (ComSoc).

Mostafa M. Fouda received his Ph.D. in Applied Information Sciences in 2011 from Tohoku University, Japan. He received B.Sc. with honors in Electrical Engineering, and M.Sc. in Electrical Communications from Benha University, Egypt, in 2002 and 2007, respectively. He is currently serving as a Postdoctoral Researcher at Department of Electrical and Computer Engineering, Tennessee Tech University, TN, USA. Also, He has been serving at Faculty of Engineering at Shoubra, Benha University, Egypt, since 2002, and in June 2016, he was promoted to the position of an Associate Professor. He was a recipient of the prestigious 1st place award during his graduation from the Faculty of Engineering at Shoubra in 2002. His research interests include Cyber Security, Internet of things, 5G communications, software-defined networks (SDNs), and smart grid communications. He is a Senior Member of the Institute of Electrical and Electronics Engineers (IEEE), and IEEE Communications Society (ComSoc).

Blockchain-Based Distributed Key Management Approach Tailored for Smart Grid

Mohamed Baza, Mostafa M. Fouda, Mahmoud Nabil, Adly Tag Eldien, Hala Mansour, and Mohamed Mahmoud

Abstract Smart grid (SG) is a new technology which enables the electrical power grid to be efficient, resilient and less pollutant. The Advanced Metering Infrastructure (AMI) is one of the key components in smart grids that enables two-way communication between end users and the utility using smart meters installed at end users. Cyber security plays a fundamental role to secure communications in the AMI. To ensure confidentiality and integrity, key management is considered a challenge in the AMI. Unfortunately, most of the existing key management schemes adopt a centralized architecture, which depends on a single entity to distribute keys and update them. In this chapter, we propose a distributed key management approach to secure communications in the SG. First, a key agreement protocol between the utility and smart meters is proposed. Then, we propose an efficient distributed multicast key management scheme so that group members can manage the group communication in a contributory way. This is attributed to blockchain technology that allows a distributed peer-to-peer network in which distrusted entities can interact with each other securely without the need to a trusted intermediary. The security and performance evaluations of our proposed approach demonstrates its effectiveness

---

M. Baza (✉) · M. Nabil · M. Mahmoud
Department of Electrical and Computer Engineering, Tennessee Tech University, Cookeville, TN, USA
e-mail: Mibaza42@students.tntech.edu; mnmahmoud42@students.tntech.edu; mmahmoud@tntech.edu

M. M. Fouda
College of Engineering, Tennessee Tech University, Cookeville, TN, USA

Faculty of Engineering at Shoubra, Benha University, Benha, Egypt
e-mail: mfouda@feng.bu.edu.eg

A. T. Eldien · H. Mansour
Department of Electrical Engineering, Faculty of Engineering at Shoubra, Benha University, Benha, Egypt
e-mail: adlytag@feng.bu.edu.eg; hala.mansour@feng.bu.edu.eg4

Z. Md. Fadlullah, A.-S. Khan Pathan (eds.), *Combating Security Challenges in the Age of Big Data*, Advanced Sciences and Technologies for Security Applications, https://doi.org/10.1007/978-3-030-35642-2_11

and scalability by taking into account the computation and communication costs which are main concerns in the big data era.

Keywords Key management · Blockchain · Smart grid · AMI networks · Cyber security

1 Introduction

The Smart Grid (SG), also known as the intelligent grid, the future grid, or the so-called intelligrid, is the convergence of information technology, communications, and power system engineering to provide a more robust, efficient and flexible electrical power system [1]. It enables two-way flows of both electricity and real-time energy-related information sent by smart devices. Some of the expected benefits of SG are preventing blackouts and providing greater availability of electricity to homes at a low cost, and also opening the door to new grid paradigms such as microgrids and Plug-in Hybrid Electric Vehicles (PHEVs) [2]. Thus, the SG paradigm will dramatically transform the state-of-the-art energy grid into a distributed power generation system which reduces the cost of the electricity generation and distribution [3].

One component of the smart Grid is Advanced Metering Infrastructure (AMI) that installs smart meters (SMs) at customer side [4]. The SMs should send fine grained power consumption readings to the utility for energy management purposes. Moreover, AMI is responsible for maintaining Demand Response (DR) tasks, which aim to reduce the consumption of electricity or shift it from on-peak to off-peak periods depending on choices of the end users. A typical AMI involves SMs, Neighbor Area Networks (NANs), wide area network infrastructure, and Meter Data Management Systems (MDMSs).

In order to secure the communication between SG entities, data confidentiality, integrity, and availability should be ensured. *Data confidentiality* ensures that the data exchanged between two entities should be protected from unauthorized access of other entities. *Data integrity* indicates that the received data (e.g., DR commands and so forth) should be similar to the transmitted data with no modifications or additions. Moreover, *availability* aims to ensure that the data is accessible and the system is available. To meet these security requirements in the smart grid, several organizations (e.g., Cyber Security Working Group (CSWG) led by National Institute of Standards and Technology (NIST) [5]) are interested in finding solutions to thwart known attacks such as man-in-the-middle (MITM) attack, Denial of Service (DoS) attack, sniffing on SMs, impersonation attack, spoofing and replay attacks [6, 7] and so forth.

To ensure both confidentiality and integrity in the SG system, the key management is a critical yet open research issue [8, 9]. The key management of the SG system needs to effectively deal with the process of key generation, establishment, storage, revocation, and update [10]. Most of the existing schemes depend on a traditional client-server model for the key management process in the SG. However, such client-server model may not be efficient in the AMI for many reasons. First, AMI has a large number of SMs, whose number may be at magnitude of millions. As a result, depending on a client server model is not scalable and may impose huge overhead on the server-end. Second, securing multicast communications needs the group key to be refreshed and redistributed (rekeying) securely whenever a group member changes to achieve forward and backward secrecy. As a result, the client-server approach poses challenges of rekeying efficiency, especially in a very scalable complex network such as AMI. Third, future SG will enable different network structures such as microgrids which can be defined as a small-scale SG system that may be isolated from the main grid and can supply the SG system with electricity. However, existing key management schemes do not consider the possibility of isolation of the microgrid.

Recently, blockchain has gained a notable attention from various fields [11, 12] since it can solve problems that have afflicted us for many years. The blockchain is a synchronized and distributed *ledger* which stores a list of blocks. No central managers are required to maintain the blockchain structure, and instead, the public ledger is secured by all the network participants. Each peer acts as a node in the network and can participate in calculating the solution to a hash-based mathematical problem to assure the integrity of transactions. Each transaction record is added to the existing block chains. All information is then updated synchronously to the entire network so that each peer keeps a record of the same ledger. Moreover, the distributed structure of blockchain can improve robustness against the single point of failure.

In this chapter, we contribute by taking into account the shortcoming of the existing key management schemes used for the SG leveraging blockchain technology. The main contributions of our work are as follows. (1) We propose an efficient end-to-end key agreement protocol, which is based on the Diffie-Hellman (DH) key establishment with less computation overhead. (2) We propose a blockchain-based self-organized or distributed multicast key management for secure communication through the SG. A one-purpose blockchain is presented to allow group members in a distributed way to securely manage the group rekeying operations whenever a meter joins or leaves the group. Then, we explain how the group members can achieve forward and backward secrecy. (3) We conduct security and performance evaluations for our proposed scheme and compare them with existing schemes to prove that our proposal is secure, efficient and scalable.

The remainder of this chapter is organized as follows. In Sect. 2, a review of the previous efforts on key management in smart grids is presented. In Sect. 3, we describe our considered system model with the relevant security considerations. In Sect. 4, we present our end-to-end key agreement protocol and a blockchain based distributed multicast key management. Then, in Sect. 5, we provide a security and performance analysis of our proposed scheme. Finally, the chapter is concluded in Sect. 6 and acknowledgment is added below the Sect. 6.

2 Literature Review

This section discusses the related works in the area of key management in smart grid. The discussion in this section is divided into two parts; key management in AMI and blockchain research on smart grid.

2.1 Key Management in AMI

Several key management schemes have been proposed for securing communications through the different elements of the smart grid. In this section, we review these efforts and critically analyze several relevant ones.

Baza et al. [13] proposed a light-weight authentication scheme based on the DH [14] protocol. This scheme provides a solution for authenticated key agreement between the Home Area Network (HAN) gateway and the Building Area Network (BAN) gateway. The objective of this scheme is to reduce the computation cost on the resource-constrained HAN gateways. However, the scheme is vulnerable to replay and MITM attacks [15, 16].

Kamto et al. [18] developed a light-weight key management protocol based on the DH key agreement [14]. The scheme achieves a low computation overhead, especially on the SMs. However, it suffers from inefficient group key management and the MITM attack in which an adversary is able to easily share a key with both the SM and gateway as discussed in [17].

Xia and Wang [19] proposed a key distribution scheme with a low computation overhead. However, it has a high communication overhead due to the large number of exchanged messages. It also does not conider the multicast key management. It has been proven that this scheme is vulnerable to MITM attack and a common type of DoS attack, called desynchronization attack [20].

Liu et al. [21] proposed a key management scheme to provide secure unicast, multicast, and broadcast communication based on the key graph technique [22]. The proposed scheme depends on generating the session keys based on previously stored keys and additional counters. The scheme uses simple cryptographic algorithms for key generation and refreshing to overcome the computation and storage constraints of SMs. However, it suffers from desynchronization attacks in which once an attacker blocks the path of data, the counters on both the SM and the management side will be different. As a consequence, both parties are unable to establish shared session key. Also, the scheme proposed in [21] suffers from inefficient group key update [10].

Inspired by the work conducted by Liu et al., another scalable key management scheme was envisioned by Wan et al. in [10]. First, Wan et al. proposed an end-to-end key establishment protocol between the SMs and the head end based on the bilinear pairing [23]. However, the end-to-end key agreement protocol suffers from a substantially high computation overhead, especially on the resource constrained

SMs due to using pairing operations. Then, Wan et al. used the session keys created by the key establishment protocol to provide a centralized multicast key management adopted from the One-way Function Tree (OFT) approach for key management [24]. Although the scheme provides less computation overhead in the rekeying process, the SM has significantly high storage overhead to maintain the binary tree in a balanced manner [25].

In [26], we introduced the idea of using self-organized or distributed multicast key management in which no central server is used to manage the group communication in the microgrids. Through public key table stored in each meter, meters can collaborate in a decentralized way to manage and update a multicast key management protocol. However, the scheme suffers from false data injection attack since a malicious meter can insert false data to the public key table stored in meters memory.

2.2 *Blockchain in SG*

The blockchain has become widely popular especially in financial applications [27]. In this section, we review some of the efforts of adopting blockchain technology in the SG. Guan et al. [28] propose a privacy-preserving and efficient data aggregation scheme based on the blockchain for smart grid communications. Instead of existing aggregation schemes that need a third-party to collect meter readings, the SG is divided into different groups, and each group has a private blockchain to record its members' data. In each time slot, a mining node is selected based on the user's electricity consumption data. Then, the mining node records all user's data into the blockchain and publishes it to other meters in the group to ensure the readings' integrity. To preserve users' privacy from other users in the same group, the scheme uses pseudonyms to hide user's identity and each user associates his data with multiple pseudonyms for further obfuscation.

In [29], Liang et al. proposed a distributed blockchain-based distributed framework for SG. The authors first point out the consequences of external cyber-attacks like false data injection attacks which may lead to wrong decisions by the control center. Then, they proposed a distributed blockchain-based data protection framework for enhancing the security of the modern power system against cyber-attack. A distributed blockchain is presented so that readings of meters are stored in immutable ledger. The simulation experiments indicate that using the blockchain in SG can efficiently protect against data manipulation attacks.

In [12], Baza et al. proposed a blockchain based firmware update for autonomous vehicles. A consortium blockchain made of different AVs manufacturers is used to ensure the authenticity and integrity of firmware updates. Instead of depending on centralized third parties to distribute the new updates, AVs, namely distributors, are allowed to participate in the distribution process by taking advantage of their mobility to guarantee high availability and fast delivery of the updates. To incentivize AVs to distribute the updates, a reward system is established that maintains a credit reputation for each distributor account in the blockchain. A zero-knowledge proof protocol is used to exchange the update in return for a proof of distribution in a trust-less environment. Moreover, attribute-based encryption (ABE) scheme is used to ensure that only authorized AVs will be able to download and use a new update.

3 System Model and Problem Formulation

In this section, we discuss the considered network model followed by pointing out the problem statement to clarify the basic requirements to secure communication within the AMI.

3.1 System Model

As shown in Fig. 1, the main components of the smart grid are the SMs, the Wide Area Network (WAN), the utility management system and the microgrids.

Smart Meters (SMs) SMs are digital devices, which are responsible for measuring electricity usage, maximum demand, current, voltage, and managing dynamic pricing policies, and remote turn on/off. Moreover, they can allow Demand Response (DR) programs [3] to increase the efficiency of SG [30]. In our scheme, we assume that meters are able to communicate with each other through wireless communication channels.

Wide Area Network (WAN) It allows the two-way communication between different SG elements (i.e., consumers) and the utility. Various communication technologies can be used such as optical fiber, power line carrier, copper [37], radio frequency cellular networks or Internet Protocol (IP)-based networks [31].

Utility Management System It includes the "brain" of the AMI. Particularly, the Meter Data Management System (MDMS) has a database of the whole system with analytical tools [10]. Furthermore, it includes an AMI Head End (AHE) to communicate with SG end users i.e., customers or consumers.

Neighborhood Area Networks (NAN) The NAN is formed by a large number of SMs in a certain neighborhood. NANs can be divided into different groups. One possible way is according to their electricity consumption type, or the coverage area. In addition, by exploiting the consumers' ability to generate electricity locally, recently, a new concept called microgrid, is gaining significant research attention. A microgrid [32] generally is defined as a low voltage network with distributed generation sources, together with local storage devices and controllable loads, e.g., water heaters and air conditioner. The microgrid may be either connected to the main power grid or in an isolated (i.e., "island") mode. In the normal situation, users

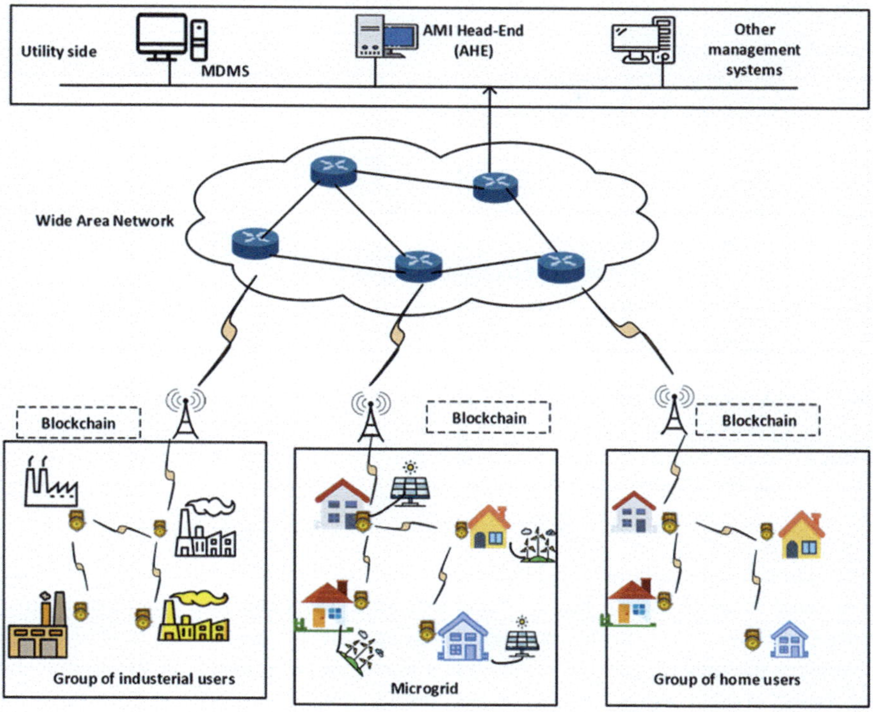

Fig. 1 Illustration for the considered network model with three different NANs. (1) A group of industrial users, (2) a group of home users and, (3) a microgrid

in the microgrid can use Distributed Energy Resources such as wind turbines, solar panels, and fuel cells to generate low voltage electricity and provide the power grid with electricity. On the other hand, it may operate in an "island" mode whereby it can still generate electricity but without exchanging electricity with the main grid [8]. Communications of the NAN can be facilitated by the Fourth Generation (4G) cellular technologies such as Long-Term Evolution (LTE), WiMAX, and so forth [33–36] as shown in Fig. 1. In addition, the notations and abbreviations that will be used in this chapter is defined in Table 1.

3.2 *Problem Formulation*

Securing the AMI is a key element to increase the reliability and efficiency of SG. Several vulnerabilities may exist in the AMI system including attacks like MITM, DoS, eavesdropping, masquerading, message replay, message modification, traffic analysis, and unauthorized access. In order to combat such malicious attacks and, at the same time, to ensure data confidentiality and integrity in the AMI, an efficient key management is required to secure communications through it. Most of

Table 1 Notations

Symbol	Description
$\oplus$	An exclusive OR operation.
n	Number of SMs in a multicast group.
H	A secure hash function.
k_G	The group key.
k_{NG}	The new group key after a member change.
k	A concatenation operator.
$\sigma_d(m)$	The signature on a message (m) using a private key d.
Kgen(1^b)	A secure b-bit key generation algorithm.
T_i	The recorded time instance of sending the message.
m_i	The message.
PKT	A public key table contains the parent binary code associated with the member public key.
$HMAC_{Ki}$	A hash-based message Authentication code (MAC) generation algorithm by using the shared session key K_i.

the existing key management schemes are centralized which rely on a single entity to manage keys. However, key management techniques can take other forms than being centralized [38]. In this chapter, we extend our previous work in [26] which introduces the idea of using distributed key management protocol in which no central server is required to secure the communication. Therefore, in the following section, we first present an end-to-end key agreement protocol that establishes a shared key between the AHE and SMs in different NANs. Secondly, we present a blockchain-based distributed multicast key management scheme tailored for securing communications within the NANs in the SG.

4 Proposed Scheme

In this section, we present our key management schemes tailored for securing communications of AMI in the SG. An end-to-end key agreement is first presented to secure unicast communication between the AHE and SMs in the NAN. Second, a blockchain based distributed key multicast management is presented to secure group communications within the NAN (Fig. 2).

4.1 End-to-End Key Establishment

Initialization Let $G = <g>$ a multiplicative group with a generator. The following preliminaries and assumptions should be considered.

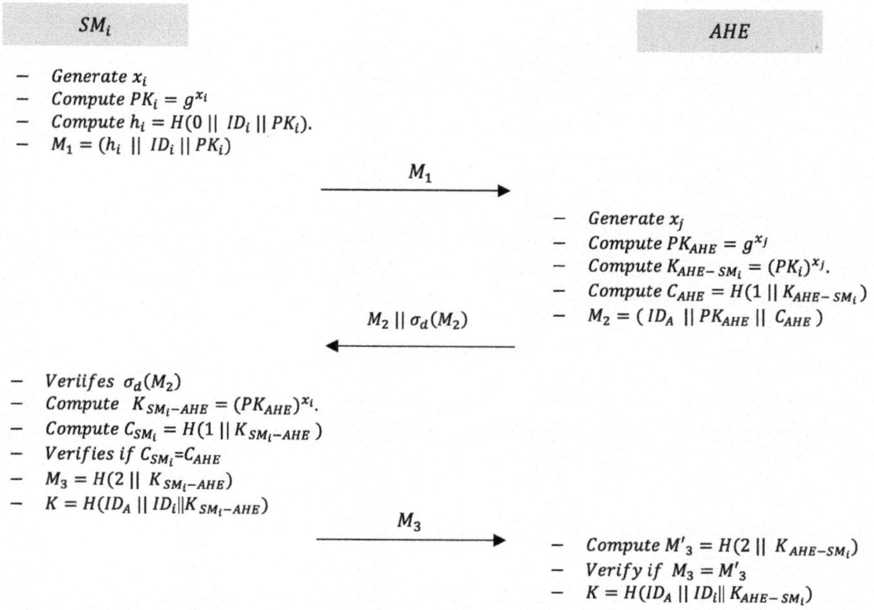

Fig. 2 Our proposed end-to-end key establishment scheme

- The discrete logarithm assumptions should hold, i.e., given g^{x_i} and g, it is computationally hard to calculate x_i given g^{x_i} . Also, the assumptions of Computational Diffie-Hellman (CDH) hold, i.e., given g^{x_a} and g^{x_b} where x_a, $x_b \in G$, it should be computationally hard to calculate $g^{x_a x_b}$.
- AMI Head-End should have a private/public key pair (e, d) which is used for the signing algorithm [39].
- Each SM should be preloaded by the parameters (g, p, e) and corresponding certificates [8].

The Proposed End-to-End Key Agreement Protocol The AHE with identity (ID_A) and smart meter SM_i with identity (ID_i) in a NAN run the following authenticated key agreement that needs only three messages.

Step 1. First, the meter SM_i chooses a random number, $x_i \in G$ to compute its public key $PK_i = g^{x_i}$ then it calculates $h_i = H(0 \| ID_i \| PK_i)$. Finally, it sends $M_1 = (h_i \| ID_i \| PK_i)$ to the AHE.

Step 2. Upon receiving the initiating message from SM_i, AHE does the following.

1. Verifies the received h_i by making sure that if $h_i = H(0 \| ID_i \| PK_i)$ where- $ID_i \| PK_i$ are the received ID_i and PK_i respectively.
2. Generates a random element $x_j \in G$ and computes $PK_i = (PK_i)^{x_j}$
3. Computes the shared key $K_{AHE-SM_i} = (PK_i)^{x_j}$.

4. Computes $C_{AHE} = H\left(1 \| K_{AHE-SM_i}\right)$.
5. Sends $M_2 = (ID_A \| PK_{AHE} \| C_{AHE})$ *as well as* $\sigma_d(M_2)$ to SM_i. Where $\sigma_d(M_2)$ is the AHE signature on M_2

Step 3. The SM verifies the signature of the AHE. Then, if the verification is "ok", it calculates $K_{SM_i-AHE} = \left(PK_{AHE}\right)^{x_i}$. After that, it computes $C_{SM_i} = H\left(1 \| K_{SM_i-AHE}\right)$. It checks the integrity of C_{AHE} by determining whether C_{SM_i} is equal to received C_{AHE} or not. At the end, it calculate $K = H(ID_A \| ID_i \| K_{SM_i-AHE})$ as the shared session key with the AHE, and then computes $M_3 = H\left(2 \| K_{SM_i-AHE}\right)$ and sends it to AHE.

Step 4. Upon receiving the message M_3, AHE computes $M'_3 = H\left(2 \| K_{AHE-SM_i}\right)$ and checks if $M_3 = M'_3$, if it is "ok", this acts as a key confirmation acknowledgment and then it sets $K = H(ID_A \| ID_i \| K_{AHE-SM_i})$ as the shared se session key.

4.2 Blockchain-Based Multicast Key Management

For efficient management of multicast keys in the AMI, the end-to-end key obtained in the above protocol is integrated with the key tree technique as well as the blockchain technology. We adopt the key tree technique scheme in [40] to be employed with a group of SMs. Indeed, the scheme is based on the one-way function tree technique [24] in which meters are organized in a binary tree as depicted in Fig. 3 where each node has two codes as follows.

1. A binary code is used to discover the position of the member. A decimal code is used to compute the intermediate nodes' key(s). Any meter can calculate all decimal codes which belongs to it by removing the last digit from the right. For instance, in Fig. 3, if SM_1 has a decimal code (045) as its decimal parent code, the meter can remove the last digit to get (04) as the decimal code for (00) node.

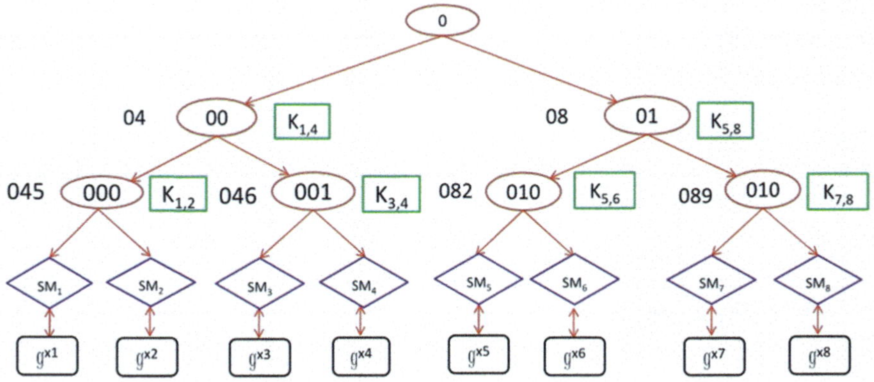

Fig. 3 A group of eight SMs that consists a multicast group which is organized in a binary tree. Each node has two codes, i.e., a decimal one and a binary one. Also, the intermediate codes are shown in green boxes

2. A decimal code is used to compute the intermediate nodes' keys (K_I) which are used to encrypt data to specific meters (multicast communications) within the group. The intermediate node key K_I is calculated by the following formula:

$$K_I = H\left(K_G \oplus Code_I\right)$$

Where K_G is the group key and $Code_I$ the decimal code that is associated with the binary node and can be calculated by concatenating the decimal code of its parent node with a random digit as illustrated by following expression,

$$Code_{Child} = Code_{Parent} \oplus Random\,number$$

As an example, in Fig. 3, SM_1 has a parent code (000) and a decimal code (045). By removing the last digit in 045, SM_1 can obtain (04) as the decimal code for the binary node (00). It also can determine its intermediate node keys as follows.

$$K_{1,2} = H\left(K_G \oplus 045\right) \text{and}\, K_{1,4} = H\left(K_G \oplus 04\right)$$

Where: $K_{1,\,2}$ can be used to encrypt group key communication between SM_1 and SM_2. Also, $K_{1,\,4}$ can be used to encrypt data between SM_1, SM_2, SM_3 and SM_4.

Besides, each meter is equipped with specific software to support the generation of a public key and private key. Finally, each meter should maintain Public Key Table (PKT) in its memory. Indeed, PKT is used to store the public key of SMs associated with their parent binary codes. Meters can use PKT to share a secret session key between group members. The contents of the PKT is updated once a meter joins or leaves a multi-cast group. To ensure integrity of keys of PKT, the keys are stored in a blockchain.

Blockchain Construction To guarantee the integrity of the PKT since there is no central server to execute the key management process, we propose the use of blockchain technology to allow SMs in a decentralized way to keep records of the PKT so that once any meter joins or leaves the group, the PKT is updated in the blockchain. Any change to the PKT is stored in a ledger formed by chronologically connected blocks. Construction of the proposed blockchain is shown in Fig. 4.

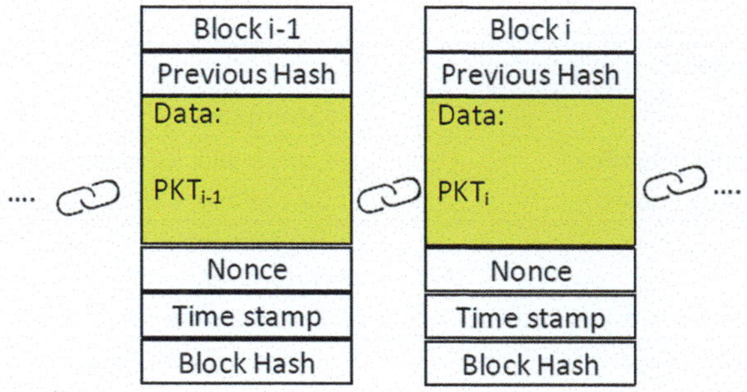

Fig. 4 Distributed ledger structure

In the following, we discuss in detail construction of the one-purpose blockchain used in our scheme, which consists of the following steps:

1. *Data Broadcast and Verification.* In this phase, once a SM joins or leaves the group, the PKT should be updated and stored on the blockchain. The joined smart meter should broadcast its newly public key signed with its private key to all meters. All meters which receive the broadcast information need to verify the received message signature. To achieve consensus on the received message before a verified data is added to a block, we adopt an address-based distributed voting mechanism [29]. The basic idea is that each meter has one chance to verify the authenticity of the received message. The voting scheme works as follows, assume K meter in a NAN where each one votes on the verification result for the received message. The data is accepted only when the following condition is true:

$$\frac{N}{K} > T \tag{1}$$

Where N represents the number of most votes and T is a threshold whose value should be greater than 50% so that meters can make sure that the voting result on certain message has been approved by the majority of meters.

2. *Mining and Generation of Blocks.* The data in the blockchain is stored through chain of linked blocks. In the blockchain network, each block has the following contents: block number, data content, timestamp, previous block hash, block hash result, and nonce solution [41, 42]. The descriptions of the contents are given in Table. 2. As in Bitcoin [27], SHA-256 is used to mine the block to obtain a hash value with certain criterion, using data content on the $(i-1)$-th block and the current timestamp, some meters can solve a puzzle to find a certain nonce value. Let C_i be the current block content as follows:

$$C_i = B || D || t_s || H(C_{i-1}) || \; nonce \tag{2}$$

Where, B is block number, D is the updated PKT data, t_s, C_{i-1} is the previous block content and *nonce* which is a random number. SHA-256 is applied to the

Table 2 Block contents

Item	Description
Block number	Current block sequence number.
Data content	Current block PKT encapsulated data.
Timestamp	The time when the last verified PKT data is added into the current block
Previous block hash	The hash value of the previous block.
Block hash	A hash value of the current block.
Nonce solution	Puzzle solution for the current block.

block content C_i to find a certain *nonce* value such that *Digest* is less than a certain target value, $\mathcal{T}$ as:

$$Digest \leq \mathcal{T} \tag{3}$$

To solve that puzzle, there is no way other than trying different values till the required $\mathcal{T}$ is obtained. The difficulty of solving the puzzle depends on the value of $\mathcal{T}$. The lower the target value, the higher the difficulty of finding solution for the puzzle. Any meter in the network can work as a miner to find the puzzle's solution. After finding the nonce, it will be broadcasted to the remaining meters so they can check if (3) applies. Then, the address-based voting scheme in (1) is used to vote on the puzzle's solution. The voting is required to ensure that majority of meters agreed on the received nonce solution. Once, the two conditions are satisfied, the block content can be added to the ledger. The major challenge is the selection of the meter who are able to solve the puzzle problem in the mining process. One possible way is that some pre-determined meters should have high computational resources to be able to act as miners, and are responsible for solving the puzzle problem. The main advantage of that solution is its low deployment cost. Another solution is that all meters have the same computational resources and miners are selected on a random basis. However, this solution is more complex and costlier since it requires upgrading the hardware of current meters.

Smart Meter Joining When a new SM joins the group, the joining protocol is executed to update the group key and the intermediate node codes. The SM joining protocol is illustrated in Fig. 5. First, when the new meter SM_8 joins the group, it

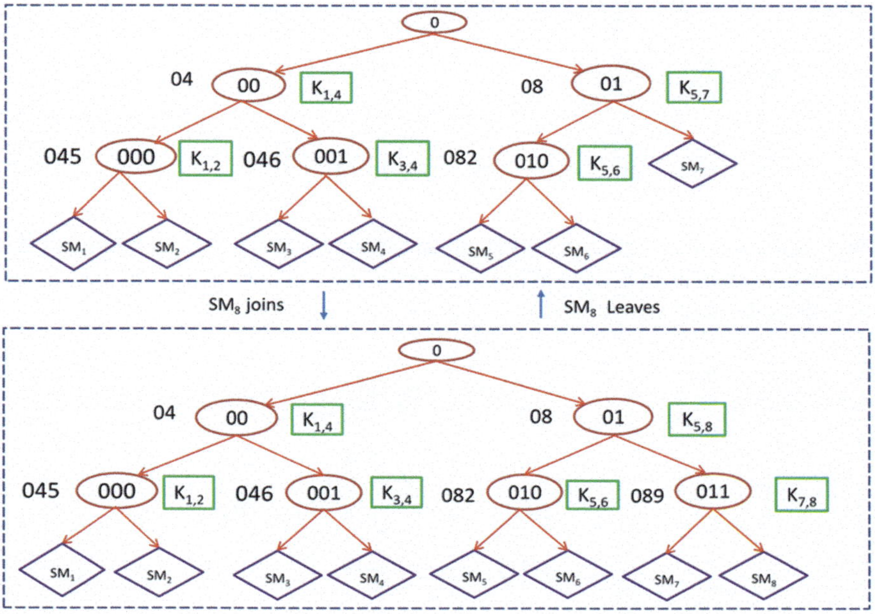

Fig. 5 Smart meter joining and eviction

should select a random element and $x_8 \in Z_p^*$ and calculates g^{x_8}. Then, it computes a signature on g^{x_8} using its private key. Then, it broadcasts g^{x_8} as well as its signature to other meters. All meters which receive the broadcast message to verify the received message signature. Then, the result is tested by the address-based distributed voting mechanism in Eq. 1, which ensures if there are enough meters agreeing on the received g^{x_8} value. Note that the voting is required to ensure that the joining meter SM_8 does not pollute the PKT by sending different values for g^{x_8}. Then, some meters can operate as miners by attempting to solve the puzzle problem in Eq. 3. Once the first meter gets the nonce value, it broadcasts the nonce to other meters so that they can verify whether the solution is valid by checking whether it satisfies the condition in 4 or not. The address- based distributed voting mechanism is utilized again to vote on the verification result before the current PKT block to be allowed to cryptographically connect to the previous ledger (Fig. 6).

Second, the meter, SM_7, which has no sibling, updates its position from (01) to (011), and then computes a decimal code associated with its parent code (011) as in Eq. 4. After that, it uses the PKT to share a shared session key (K_s) with SM_8 as in 5). Then, it updates the group key by applying a one-way function to the previous group key as in (6) and sends them to SM_8 encrypted by the shared session key. Also, we employ a hash-based Message Authentication Code (MAC) generation algorithm by using K_s on the sent message to ensure integrity as in (7).

$$code_{node_{011}} = (08 \| 9) = 089 \tag{4}$$

$$SM_7 : K_s = H\left(\left(g^{x_8}\right)^{x_7}\right),\ SM_8 : K_s = H\left(\left(g^{x_7}\right)^{x_8}\right) \tag{5}$$

$$K_{NG} = H(K_G) \tag{6}$$

$$SM_7 \rightarrow SM_8 : \{K_{NG}, 089, HMAC_{K_S}\}_{Encr_{K_S}} \tag{7}$$

The other remaining group members update their group key by just applying the one-way function as in Eq. (3). Also, SMs in the affected path renew the intermediate node codes as illustrated in Eqs. (8) and (9).

Fig. 6 PKT is updated when *SM*8 joins

Node	Public Key
000	g^{x_1}, g^{x_2}
001	g^{x_3}, g^{x_4}
010	g^{x_5}, g^{x_6}
011	g^{x_7}

⇒

Node	Public Key
000	g^{x_1}, g^{x_2}
001	g^{x_3}, g^{x_4}
010	g^{x_5}, g^{x_6}
011	g^{x_7}, g^{x_8}

$$SM_7, SM_8 \leftrightarrow K_{7,8} = H\left(K_{NG} \oplus 089\right) \tag{8}$$

$$SM_5, \ldots, SM_8 \leftrightarrow K_{5,8} = H\left(K_{NG} \oplus 08\right) \tag{9}$$

Smart Meter Eviction When a SM leaves a group (e.g., when it is defected/encounters malfunctions, or requires maintenance), the forward secrecy needs to be ensured. In this vein, the group key needs to be renewed using the following SM eviction protocol. If SM_8 is to leave the group, its sibling (SM_7) upgrades its position in the tree to (01) and updates the PKT by deleting the leaving meter's associated key as shown in Fig. 7. Also, it informs the other SMs about the change by sending a broadcast message. Note that the same procedures regarding updating the PKT in the blockchain is followed as in Sec. 4.2. This is mandatory to ensure the integrity of the PKT in a distributed network. Then, SM_7 computes a new group key (K_{NG}) by employing the Kgen(1^b) algorithm and uses its PKT to share a key with any SM in each branch of the tree. Then, it sends the new group key to them as illustrated in (10) and (11).

$$SM_7 \rightarrow SM_1 : \left\{K_{NG} \oplus HMAC_{K1}\right\}_{Encr_{K1}} \tag{10}$$

$$SM_7 \rightarrow SM_5 : \left\{K_{NG} \oplus HMAC_{K2}\right\}_{Encr_{K2}} \tag{11}$$

Where $K_1 = H\left(g^{x_1}\right)^{x_7}$ and $K_2 = H\left(g^{x_5}\right)^{x_7}$ are shared secret keys between (SM_7 and SM_1) and (SM_7 and SM_5 respectively. On the other hand, SM_1 and SM_5 can make use of the intermediate keys to send encrypted to other SMs in the group as in 13 and 14.

$$SM_1 \rightarrow SM_2, \ldots, SM_4 : \left\{K_{NG} \oplus HMAC_{K1,4}\right\}_{Encr_{K1,4}} \tag{12}$$

$$SM_5 \rightarrow SM_6 : \left\{K_{NG} \oplus HMAC_{K5,6}\right\}_{Encr_{K2}} \tag{13}$$

In order to ensure the forward secrecy, the SMs in the affected branch should update their intermediate node codes as in (14).

$$SM_5, SM_6, SM_7 \leftrightarrow K_{5,7} = H\left(K_{NG} \oplus 08\right) \tag{14}$$

Fig. 7 PKT is updated when *SM8* leaves

Node	Public Key
000	g^{x_1}, g^{x_2}
001	g^{x_3}, g^{x_4}
010	g^{x_5}, g^{x_6}
011	g^{x_7}, g^{x_8}

⇒

Node	Public Key
000	g^{x_1}, g^{x_2}
001	g^{x_3}, g^{x_4}
010	g^{x_5}, g^{x_6}
011	g^{x_7}

Secure Unicast and Multicast Communication Our scheme adopts the following message transmission methods in order to achieve confidentiality and integrity.

Secure Unicast Communications When the AHE wants to send a message (m_i) to SM_i, both of them can use the key agreement protocol discussed earlier to generate a shared secret session key (K_s) and send the following message,

$$AHE \rightarrow SM_i : \{m_i, T_i, HMAC_{Ks}\}_{Encr_{Ks}}$$

Where T_i is the recorded time stamp of sending the message in order to prevent replay attacks. Also, *HMAC* is based on the MAC of (m_i), and is used to ensure the integrity of the message.

Secure Multicast Communications The group key K_G is used to secure the multicast communication. As discussed earlier, our multicast management approach is used to maintain K_G in a self-organized manner so that the group key can be sent by any group member to the AHE. To illustrate the multicast communication, the AHE or a meter SM_i can send a multicast message (m_i) to other meters in the group as follows.

$$AHE / SM_i \rightarrow SM_* : \{m_i, T_i, HMAC_{KG}\}_{Encr_{KG}}$$

Where, SM_* refers to all other meters. Also, *HMAC* and T_i are used to ensure the integrity of the message and prevent replay attacks, respectively.

5 Security Analysis and Performance Analysis

In this section, the security features of our proposal are discussed, and then, we evaluate the performance of our scheme using real devices.

5.1 Security Analysis

In this subsection, we analyze the security features of the end-to-end key establishment protocol first, and then we analyze the blockchain based multicast key management protocol.

Analysis of the Proposed Key Agreement Protocol Our proposed key agreement protocol is able to achieve the following security features.

Resistance to the MITM and Replay Attacks In the MITM attack, the adversary secretly relays and probably changes the communication between two participants so that they believe that they are communicating directly, while they are talking through the attacker. Also, replay attacks is by resending the same message so the attacker can establish a session key with the AHE. In our end-to-end key agreement, we allow only legitimate meters to share a shared secret keys with the AHE. An adversary who want to deceive the meter and share a key with him, should compromise the AHE private keys. To further illustrate, upon receiving M_2 from the AHE, the meter verifies the AHE signature on it. That ensures only AHE to establish shared keys with the meters. Then, to confirm the shared key, the meter computes M_3, which includes the number two and sends it to the AHE who in turn checks the message M_3. The role of M_3 to act as a key management confirmation message and thwart the replay attack since if an attacker try to reply the message M_1 to establish the same share key with the AHE, that would be hard due to the incremental numbers used through the protocol. In this fashion, the MITM attack scenarios can be thwarted by adopting our proposal.

Data Integrity Data integrity can be ensured in our scheme because of the implemented hash function (H) as follows consider a situation where SM_i sends its identity ID_i and the public key (PK_i) concatenated with the hash value of both. Upon receiving the message, the AHE can verify the integrity of the received message by recalculating its hash value. In our protocol, SM_1can be verified by verifying and h_1 is ensured by the AHE signature and M_3 is ensured by the established shared secret key between AHE and SM_i.

The Blockchain Based Multicast Key Management Protocol The proposed multicast key management protocol has the following security properties.

Distributed Feature Based on the Blockchain Technology Our proposed multicast key management scheme removes the need for a centralized entity to generate or update security keys while facilitating the re-keying process. Moreover, the blockchain network can protect against manipulation attacks of the public keys. It is required to ensure the integrity of the PKT allowing meters in a distributed way to collaboratively update their group keys whenever a meter joins or leaves the group. If an external attacker (who is not eligible to join the group e.g., not having a valid certificates [8, 16]) wants to join the multicast group, he should manipulate the majority of nodes by compromising their private keys. However, taking control of the majority of meters to produce a false agreement on the voting result is hard to achieve. To further illustrate, consider a NAN with 1000 meters and the threshold according to Eq. 1 is set to be 60%. Then, to achieve false agreement on the voting result, an attacker needs to control at least 600 meters simultaneously which is considered completely hard.

Backward Secrecy Our proposal achieves backward secrecy to prevent a new member from decoding exchanged messages prior to the actual joining time. As illustrated in the joining process, when SM_8 joins the group, the group key is renewed and the intermediate node codes in the affected path should be updated. By doing so, the SM is not able to decipher even the previous messages.

Forward Secrecy Our proposal achieves forward secrecy to prevent an evicted or removed group member from continuing to access the communication within the group. This can be achieved as follows. When SM_8 leaves the group, the tree, and the PKT are updated so that it is not able to further decrypt group messages encrypted with the new group key after it has left the group.

5.2 Performance Evaluations

In order to evaluate the performance of our proposed scheme, we consider two aspects, namely, computation and communication cost. Then, we compare these results with two relevant, existing schemes, which were proposed by Liu et al. [21] and Wan et al. [10], respectively. Since a comparable distributed key management protocol does not exist in the literature, we compare our proposal with these two schemes to demonstrate the effectiveness of our scheme, particularly when dealing with SMs.

Computation Cost Computation cost is the processing overhead needed by the meters in the key management scheme. The computation costs of the end-to-end key establishment and the multicast key management are shown in Table 3. The results for both Liu et al.'s and Wan et al.'s schemes are readily obtained from [10]. For the end-to-end key agreement, in our scheme, each SM needs to compute two exponentiation operations, one signature verification operation and four hashes. Meanwhile, the head end needs two exponentiation operations, one digital signature operation, and three hashes.

Table 3 Computation Cost Comparison

		Liu et al. [21]	Wan et al. [10]	Our Proposed Scheme
End-to-end key	AHE	–	$n(2C_P + C_M)$	$2C_{exp} + C_{sig} + 3C_H$
Establishment	SM	–	$3C_M$	$2Cexp + Cver + 4CH$
Adding a member	AHE	$(4n + 5)C_f + (n + 2)CE$	$Cf + Cr$	0
	SM	$4nC_f + nCE$	C_f	$2nC_f$
Evicting a member	AHE	$(4n + 5)C_f + (n + 2)CE$	$h(C_E + 2C_f)$ $h(C_E$	0
	SM	$4nC_f + nCE$	$+C_f)$	$2nC_E - 4C_f$

n is the number of SMs in the group, h is height of the binary tree. C_P, C_M, $Csig$, C_{ver}, C_H, C_f and C_E are, respectively, the computational cost for the bilinear pairing, the point multiplication, the RSA signing function, the RSA signature verification, the hash function H, one-way function calculation and the encryption function E

Table 4 Computation time for cryptographic operations

Cryptographic operation	MICAZ	Pentium IV 3GHz
Paring operation	5.32 s	3.88 ms
Point multiplication	2.45 s	1.82 ms
Hash function	0.023 ms	≈ 0 ms
AES encryption/decryption	0.023 ms	≈ 0 ms
Public decryption/sign(.)	21 s	16 ms

For the multicast key management, we analyze the computation cost required by SMs. Three criteria are used to estimate the overall computation cost needed by group members which are cost due to group key update, encryption overhead, and intermediate node keys update [10, 40] as illustrated below.

- For the join operation: Each SM needs to compute one C_f to get the new group key that totally costs nC_f and two C_E encryption overhead. Also, the computation cost due to the update of intermediate node codes can be approximated to $(n-2)C_f$. Hence, the total computation overhead $= n\,C_f + (n-2)\,C_f + 2\,CE = 2nC_f$.
- For the departure operation: Only one C_f is required for the new group key generation. Then, each SM should perform at least one encryption operation to get the group key which costs nCE. Also, the computation cost due to the update of the intermediate node codes can be approximated to $(n-5)\,C_f$. Therefore, the total computation overhead is equal to $C_f + nCE + (n-5)\,C_f = 2nCE - 4C_f$.

In order to simulate our scheme on a real environment, cryptographic operations are carried out in wireless sensor nodes such as MICAZ sensors, which emulate SMs in nature. Each emulated SM has 4 KB Random Access Memory (RAM), 128 KB Read Only Memory (ROM), and equipped with a low-power ATmega128L micro-controller working at 7.3 MHz. Also, a 3GHz Pentium IV PC is set up as the head end. We assume that the encryption is conducted by employing the Advanced Encryption

Standard (AES) with a 128-bit symmetric key. Table 4 lists the estimated time for performing the various cryptographic operations (adopted from [39, 43]). For this particular performance evaluation of our proposed scheme, we conduct simulations based on MATLAB [44]. The simulation results are depicted in Figs. 8, 9a and b. First, for the end-to-end key agreement protocol, from Fig. 8, it can be noticed that the computation overhead on the SMs in our proposed scheme is less than that in Wan et al.'s scheme. Wan et al.'s scheme has less overhead on the head end which is a server with high computation capabilities.

However, it performs rather poorly on the SMs which typically have low computation capabilities. For the multicast key management protocol as shown in Fig. 9a and Fig. 9b, we can conclude that our scheme exhibits a much lower computation overhead on the SMs in the rekeying process than Liu et al.'s scheme. Even though Wan et al.'s scheme shows a better efficiency, our scheme has no overhead on the centralized node due to the decentralized nature of our proposed scheme. As a result, our scheme can achieve scalability and efficiency despite its distributed nature.

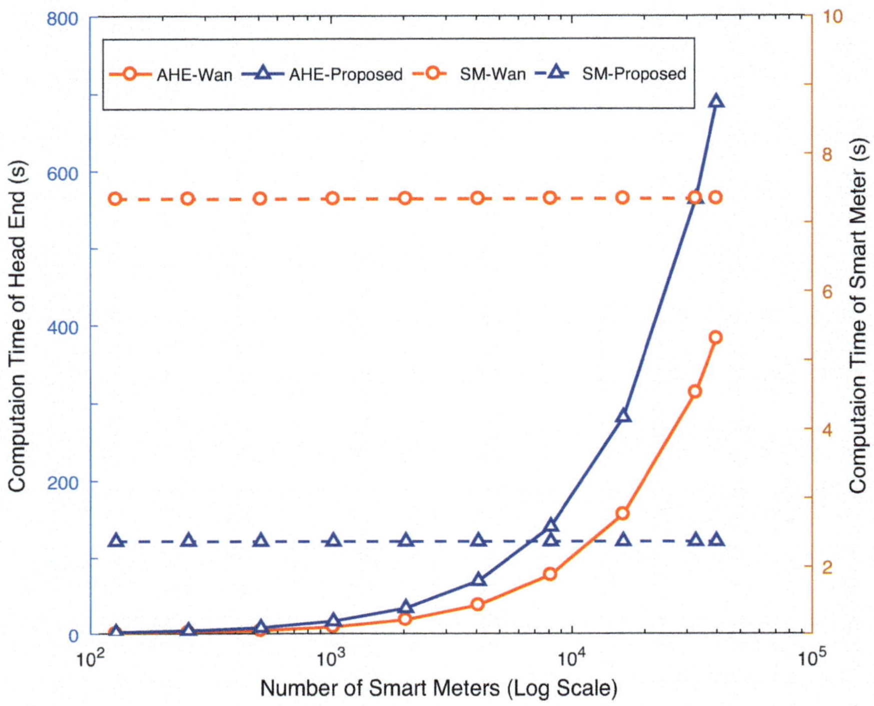

Fig. 8 Simulation result for the proposed key establishment protocol comparing with Wan et al's scheme. AHE-Wan and SM-Wan refer to the computation costs of the AHE and SM in Wan et al.'s scheme, respectively

Communication Overhead Communication overhead is measured by the number of messages required to be transmitted [45–50] in the key agreement protocol or during the group re-keying process when a member joins/leaves the group. Communication overhead results are given in Table 5. As indicated in Table 5, our scheme costs only three unicast messages to run the end-to-end key establishment which is the same as that achieved by Wan et al.'s scheme. Moreover, for the rekeying process, in our distributed multicast key management scheme, it only requires a single message to be delivered to the new member at the and one broadcast message to update the PKT by the public value of the newly added meter. Also, the eviction phase needs a number of messages of the order of $log_2 n$ as the unicast messages to renew the PKT. These results demonstrate that our scheme incurs a low communication overhead compared to that achieved by Liu et al.'s scheme and is as efficient as Wan et al.'s scheme.

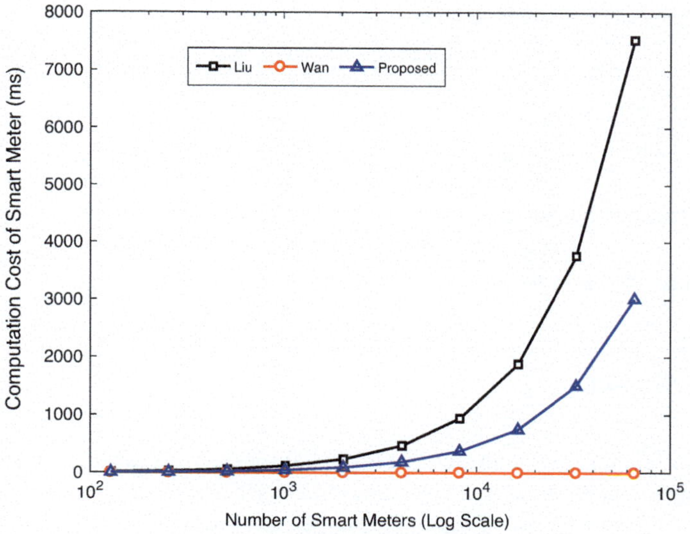

(a) Overhead of the join protocol.

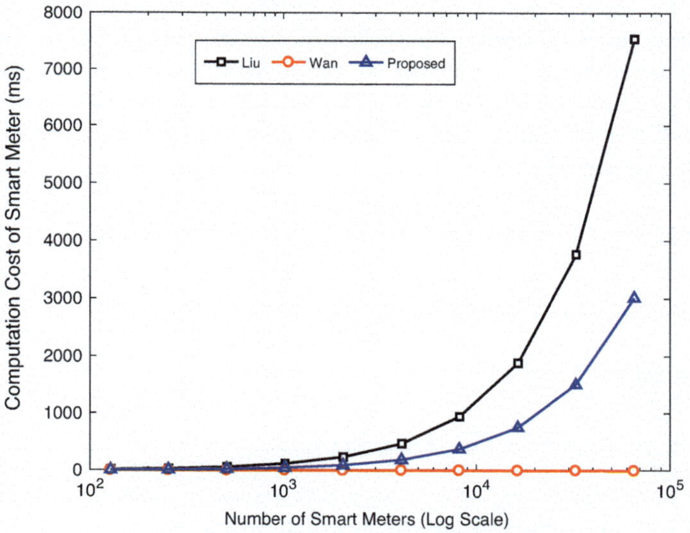

(b) Overhead of the eviction protocol.

Fig. 9 Computation cost comparison on SMs for joining or departing events of a meter

Table 5 Communication overhead comparison

		Liu et al. [21]	Wan et al. [10]	Our Proposed Scheme
End-to-end key	Broadcast	–	0	0
Establishment	Unicast	–	$3nK$	$3nK$
Adding a member	Broadcast	0	h	K
	Unicast	$2nK$	K	K
Evicting a member	Broadcast	0	$hK + h$	0
	Unicast	$2nK$	0	$2(h - 1)K$

n is the number of SMs in the group and K is the size of a key (in bits)

6 Conclusion

Secure and efficient key management is crucial for ensuring security in smart grid. In this chapter, we introduced a key management scheme to secure communications with in the AMI of the SG. First, a key agreement protocol between the AHE and SMs was proposed. Then, to secure group communication, a blockchain based multicast key management was introduced to secure group communications in which no central server needs to be used to distribute keys or update them when a member's status changes. Also, we conducted the performance evaluations of our scheme which demonstrates that also it has low computation and communication overhead on the smart meters. Our security analysis demonstrates that our scheme achieves forward and backward secrecy.

Acknowledgments This work is supported by the U.S. National Science Foundation under Grant CNS-1619250.

References

1. Fadlullah ZM, Fouda MM, Kato N, Takeuchi A, Iwasaki N, Nozaki Y (2011) Toward intelligent machine-to-machine communications in smart grid. IEEE Commun Mag 49(4):60–65
2. Akkaya K, Rabieh K, Mahmoud M, Tonya S (2015) Customized certificate revocation lists for IEEE 802.11s based smart grid AMI networks. IEEE Trans Smart Grid (TSG) 6(5):2366–2374
3. Badsha S, Yi X, Khalil I, Liu D, Nepal S, Bertino E, Lam K (2018) Privacy-preserving location-aware personalized web service recommendations. IEEE Transactions on Services Computing
4. Baza M, Nabil M, Bewermeier N, Fidan K, Mahmoud M, Abdallah M (2019) Detecting sybil attacks using proofs of work and location in vanets. arXiv preprint arXiv:1904.05845
5. Badsha S, Yi X, Khalil I (2016) A practical privacy-preserving recommender system. Data Sci Eng 1(3):161–177
6. Vakilinia I, Badsha S, Sengupta S (2018) Crowdfunding the insurance of a cyber-product using blockchain. In: Ubiquitous Computing, Electronics and Mobile Communication Conference (UEMCON), 2018
7. Al Amiri W, Baza M, Banawan K, Mahmoud M, Alasmary W, Akkaya K (2020) Towards secure smart parking system using blockchain technology. In: 17th annual Consumer Communications & Networking Conference (CCNC), IEEE, Las Vegas, USA

8. Shafee A, Baza M, Talbert DA, M. M. Fouda, Nabil M, Mahmoud M (2019) Mimic learning to generate a shareable network intrusion detection model, arXiv preprint arXiv: 1905.00919
9. Baza M, Lasla N, Mahmoud M, Srivasta G, Abdallah M (2019) B-Ride: ride sharing with privacy-preservation, trust and fair payment atop public blockchain. arXiv preprint arXiv:1906.09968
10. Wan Z, Wang G, Yang Y, Shi S (2014) Skm: scalable key management for advanced metering infrastructure in smart grids. IEEE Trans Ind Electron 61(12):7055–7066
11. Eldosouky A, Saad W, Mandayam N (2017) Resilient critical infrastructure: Bayesian network analysis and contract-basedoptimization, arXiv preprint arXiv:1709.00303
12. Baza M, Nabil M, Lasla N, Fidan K, Mahmoud M, Abdallah M (2018) Blockchain-based firmware update tailored for autonomous vehicles, arXiv preprint arXiv:1811.05905
13. Baza M, Mahmoud M, Srivastava, G, Alasmary W, Younis M (2020) A light blockchain-powered privacy-preserving organization scheme for ride sharing services. In: 91th Vehicular Technology Conference (VTC-Spring), IEEE, Piscataway
14. Diffie W, Hellman M (1976) New directions in cryptography. IEEE Trans Inf Theory 22(6):644–654
15. Baza M, Nabil M, Ismail M, Mahmoud M, Serpedin E, Rahman M (2019) Blockchain-based charging coordination mechanism for smart grid energy storage units, arXiv preprint arXiv:1811.02001
16. Baza M, Pazos-Revilla M, Nabil M, Sherif A, Mahmoud M, Alasmary W (2019) Privacy-preserving and collusion-resistant charging coordination schemes for smart grid, arXiv preprint arXiv:1905.04666
17. Mohammadali A, Haghighi MS, Tadayon MH, Nodooshan AM (2016) A novel identity-based key establishment method for advanced metering infrastructure in smart grid. IEEE Trans Smart Grid 9:2834–2842
18. Kamto J, Qian L, Fuller J, Attia J (2011) Light-weight key distribution and management for advanced metering infrastructure. In: GLOBECOM workshops (GC Wkshps), 2011 IEEE. IEEE, Piscataway, pp 1216–1220
19. Xia J, Wang Y (2012) Secure key distribution for the smart grid. IEEE Trans Smart Grid 3(3):1437–1443
20. Nicanfar H, Jokar P, Beznosov K, Leung VC (2014) Efficient authentication and key management mechanisms for smart grid communications. IEEE Syst J 8(2):629–640
21. Liu N, Chen J, Zhu L, Zhang J, He Y (2013) A key management scheme for secure communications of advanced metering infrastructure in smart grid. IEEE Trans Ind Electron 60(10):4746–4756
22. Wong CK, Gouda M, Lam SS (2000) Secure group communications using key graphs. IEEE/ACM Trans Networking (TON) 8(1):16–30
23. Chen L, Kudla C (2003) Identity based authenticated key agreement protocols from pairings. In: Computer security foundations workshop, 2003. Proceedings. 16th IEEE. IEEE, pp 219–233
24. Sherman AT, McGrew DA (2003) Key establishment in large dynamic groups using one-way function trees. IEEE Trans Softw Eng 29(5):444–458
25. Benmalek M, Challal Y (2015) Eskami: efficient and scalable multi-group key management for advanced metering infrastructure in smart grid. In: Trustcom/BigDataSE/ISPA, 2015 IEEE, vol 1. IEEE, Los Alamitos, pp 782–789
26. Baza M, Fouda MM, Eldien AST, Mansour HA (2015) An efficient distributed approach for key management in microgrids. In: Computer engineering conference (ICENCO), 2015 11th international. IEEE, Piscataway, pp 19–24
27. Al Amiri W, Baza M, Banawan K, Mahmoud M, Alasmary W, Akkaya K (2019) Privacy-preserving smart parking system using blockchain and private information retrieval, arXiv preprint arXiv:1904.09703

28. Guan Z, Si G, Zhang X, Wu L, Guizani N, Du X, Ma Y (2018) Privacy-preserving and efficient aggregation based on blockchain for power grid communications in smart communities, arXiv preprint arXiv:1806.01056
29. Liang G, Weller SR, Luo F, Zhao J, Dong ZY (2018) Distributed blockchain-based data protection framework for modern power systems against cyber attacks. IEEE Trans Smart Grid 10:3162–3173
30. Baza M, Salazar A, Mahmoud, M, Abdallah M, Akkaya, K (2020) On sharing models instead of data using mimic learning for smart health applications. In: Proceedings od IEEE International Conference on Informatics, IoT, and Enabling Technologies (ICIoT'20), IEEE, Doha, Qatar
31. Amer R, Saad W, ElSawy H, Butt M, Marchetti N (2018) Caching to the Sky: Performance Analysis of Cache-Assisted CoMP for Cellular-Connected UAVs. arXiv preprint arXiv:1811.11098
32. Lasseter RH, Paigi P (2004) Microgrid: a conceptual solution. In: Proceedings of power electronics specialists conference, 2004. PESC04. 2004 IEEE 35th annual, vol 6. IEEE, Piscataway, pp 4285–4290
33. Amer R, Butt MM, Bennis M, Marchetti N (2017) Delay analysis for wireless d2d caching with inter-cluster cooperation. In: Proceedings of IEEE GLOBECOM 2017–2017 IEEE global communications conference. IEEE, pp 1–7
34. Amer R, El Sawy H, Kibiłda J, Butt MM, Marchetti N (2018) Cooperative transmission and probabilistic caching for clustered d2d networks, arXiv preprint arXiv:1811.11099
35. Amer R, Butt MM, El Sawy H, Bennis M, Kibiłda J, Marchetti N (2018) On minimizing energy consumption for d2d clustered caching networks, arXiv preprint arXiv:1808.03050
36. Amer R, Elsawy H, Butt MM, Jorswieck EA, Bennis M, Marchetti N (2018) Optimizing joint probabilistic caching and communication for clustered d2d networks, arXiv preprint arXiv:1810.05510
37. Liu D, Alahmadi A, Ni J, Lin X et al (2019) Anonymous reputation system for iiot-enabled retail marketing atop pos blockchain. IEEE Trans Ind Inf 15:3527–3537
38. Rafaeli S, Hutchison D (2003) A survey of key management for secure group communication. ACM Comput Surv (CSUR) 35(3):309–329
39. Eldosouky A, Saad W (2018) On the cybersecurity of m-health iot systems with led bitslice implementation. In: 2018 IEEE international conference on consumer electronics (ICCE). IEEE, Piscataway, pp 1–6
40. Mortazavi SA, Pour AN, Kato T (2011) An efficient distributed group key management using hierarchical approach with diffie-hellman and symmetric algorithm: Dhsa. In: Computer networks and distributed systems (CNDS), 2011 international symposium on. IEEE, Piscataway, pp 49–54
41. Li M, Weng J, Yang A, Lu W, Zhang Y, Hou L, Liu JN, Xiang Y, Deng R (2018) Crowdbc: a blockchain-based decentralized framework for crowdsourcing, In: IEEE transactions on parallel and distributed systems
42. Weng J, Weng J, Zhang J, Li M, Zhang Y, Luo W (2018) Deepchain: Auditable and privacy-preserving deep learning with blockchain-based incentive. Cryptology ePrint Archive, Report 2018/679
43. Lee J, Kapitanova K, Son SH (2010) The price of security in wireless sensor networks. Comput Netw 54(17):2967–2978
44. Eldosouky A, Ferdowsi A, Saad W (2019) Drones in distress: A gametheoretic countermeasure for protecting uavs against gps spoofing, arXiv preprint arXiv:1904.11568
45. Akkaya K, Rabieh K, Mahmoud M, Tonyali S (2014) Efficient generation and distribution of CRLs for IEEE 802.11s-based Smart Grid AMI Networks. In: Proceedings of IEEE SmartGridComm'14, Italy, November, 2014
46. Elhoshy S, Ibrahim M, Ashour M, Elshabrawy T, Hammad H, Rizk MM (2016) A dimensioning framework for indoor das lte networks. In: 2016 international conference on selected topics in mobile & wireless networking (MoWNeT). IEEE, Piscataway, pp 1–8

47. Kelarev A, Yi X, Badsha S, Yang X, Rylands L, Seberry, J (2019) A multistage protocol for aggregated queries in distributed cloud databases with privacy protection. Futur Gener Comput Syst 368–380
48. Li M, Zhu L, Lin X (2018) Efficient and privacy-preserving carpooling using blockchain-assisted vehicular fog computing. IEEE Internet Things J 6:4573–4584
49. French A, Mozaffari M, Eldosouky A, Saad W (2018) Environment-aware deployment of wireless drones base stations with google earth simulator, arXiv preprint arXiv:1805.10424
50. Baza M, Baxter J, Lasla N, Mahmoud M, Abdallah M, Younis M (2020) Incentivized and secure blockchain-based firmware update and dissemination for autonomous vehicles. In: Connected and autonomous vehicles in smart cities, CRC press, 2020

Mohamed Baza is currently a Graduate Research Assistant in the Department of Electrical & Computer Engineering, Tennessee Tech. University, USA and pursuing his Ph.D. degree in the same department. He received the B.S. degree and the M.S. degree in Computer Engineering from Benha University, Egypt in 2012 and 2017, respectively. He was a recipient of the prestigious 2nd place award during his graduation in 2012. His research interests include Blockchains, cryptography and network security, smart-grid and AMI networks, and vehicular ad-hoc networks.

Mostafa M. Fouda received his Ph.D. in Applied Information Sciences in 2011 from Tohoku University, Japan. He received B.Sc. with honors in Electrical Engineering, and M.Sc. in Electrical Communications from Benha University, Egypt, in 2002 and 2007, respectively. He has been serving at Faculty of Engineering at Shoubra, Benha University, Egypt, since 2002, and in June 2016, he was promoted to the position of an Associate Professor. He also served as an Assistant Professor at Graduate School of Information Sciences (GSIS), Tohoku University, Japan, from February 2013 to March 2014. He was a recipient of the prestigious 1st place award during his graduation from the Faculty of Engineering at Shoubra in 2002. His research interests include 5G communications, software-defined net-works (SDNs), smart grid communications, cognitive radio networks, disaster resilient networks, wireless mesh networks, and network security. He has a noteworthy contribution toward the research community through his technical papers in scholarly journals, magazines, and international conferences in various areas of networking and communications. He has served as a Workshops Chair, Session Chair, Technical Program Committee (TPC) member, and designated reviewer in leading international conferences/workshops. He also served as a Guest Editor of some special issues of several top-ranked journals. He also serves as a referee of some renowned IEEE journals and magazines. He is a Senior Member of IEEE.

Mahmoud Nabil is currently a Graduate Research Assistant in the Department of Electrical & Computer Engineering, Tennessee Tech. University, USA and pursuing his Ph.D. degree in the same department. He received the B.S. degree and the M.S. degree in Computer Engineering from Cairo University, Cairo, Egypt in 2012 and 2016, respectively. His research interests include machine learning, cryptography and network security, smart-grid and AMI networks, and vehicular ad-hoc networks.

Adly S. Tag Eldien received the B.Sc. degree in Electronics and Communication, Benha University in 1984 and the M.Sc. in computer based speed control of single phase induction motor using three level PWM with harmonic elimination, Benha University, in 1989. The Ph.D. in optimal robot path control, Benha University, in 1993. He is currently serving as Associate Professor at Faculty of Engineering at Shoubra, Benha University, Egypt. His research interests include Robotics, Networks, and Communication Engineering.

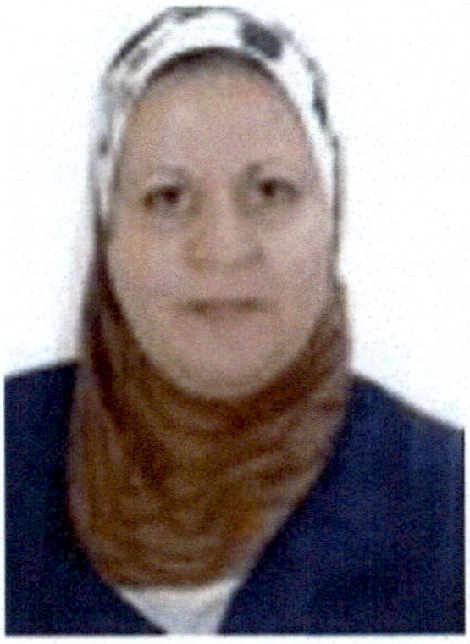

Hala A.K. Mansour is a Professor & Head of Electronics & Communication session at Faculty of Engineering at Shoubra, Benha University, where she has been since 1980. She also currently serves as Head of Credit Hours Communication & Computer Engineering Department at Faculty of Engineering at Shoubra, Benha University. She has done many investigations in the area of digital signal processing & digital design. She has more than 80 published papers.

Dr. Mohamed M. E. A. Mahmoud received PhD degree from the University of Waterloo in April 2011. From May 2011 to May 2012, he worked as a postdoctoral fellow in the Broadband Communications Research group – University of Waterloo. From August 2012 to July 2013, he worked as a visiting scholar in University of Waterloo, and a postdoctoral fellow in Ryerson University. Currently, Dr Mahmoud is an associate professor in Department Electrical and Computer Engineering, Tennessee Tech University, USA. The research interests of Dr. Mahmoud include security and privacy preserving schemes for smart grid communication network, mobile ad hoc network, sensor network, and delay-tolerant network. Dr. Mahmoud has received NSERC-PDF award. He won the Best Paper Award from IEEE International Conference on Communications (ICC'09), Dresden, Germany, 2009. Dr. Mahmoud is the author for more than twenty three papers published in major IEEE conferences and journals, such as INFOCOM conference and IEEE Transactions on Vehicular Technology, Mobile Computing, and Parallel and Distributed Systems. He serves as an Associate Editor in Springer journal of peer-to-peer networking and applications. He served as a technical program committee member for several IEEE conferences and as a reviewer for several journals and conferences such as IEEE Transactions on Vehicular Technology, IEEE Transactions on Parallel and Distributed Systems, and the journal of Peer-to-Peer Networking.

Index

Z. Md. Fadlullah, A.-S. Khan Pathan (eds.), *Combating Security Challenges in the Age of Big Data*, Advanced Sciences and Technologies for Security Applications, https://doi.org/10.1007/978-3-030-35642-2

The manufacturer's authorised representative in the EU is Springer Nature Customer Service Centre GmbH, Europaplatz 3, 69115 Heidelberg, Germany. If you have any concerns regarding our products, please contact ProductSafety@springernature.com

Printed and bound by CPI Group (UK) Ltd, Croydon, CR0 4YY
30/09/2026
02225247-0001